INTEGRATED REVIEW WORKSHEETS

ELEMENTARY STATISTICS: PICTURING THE WORLD

SEVENTH EDITION

Ron Larson

Pennsylvania State University
The Behrend College

Betsy Farber

Bucks County Community College

Reproduced by Pearson from electronic files supplied by the author.

Publishing as Pearson, 330 Hudson Street, NY NY 10013

1 18

ISBN-13: 978-0-13-476207-4
ISBN-10: 0-13-476207-X

Elementary Statistics: Picturing the World, 7th Edition

Ron Larson and Betsy Farber

Integrated Review Table of Contents

NOTE: Section reference in parentheses is the first section in the text where the corresponding skill is used.

Chapter 1 Introduction to Statistics

Reading Graphs (Section 1.1)

One way to organize data is to put it in some type of graph. This is an excellent way to transfer information in a visual way. In this section, you will read information from different types of graphs.

Bar Chart

A bar chart has horizontal bars that do not touch each other. Each bar represents a category. The categories are listed vertically. In the bar chart below, each bar represents a cycling race. This bar chart is called a stacked bar chart. Each bar represents the number of hours it took for two different racers, Racer A and Racer B, to complete a cycling race. The number of hours are listed horizontally. You can use the horizontal line to estimate each racer's time. In the bar chart below, each racer's time is stated in the bar.

The title of the bar chart is "Cycling Times." The title of the vertical axis is "Cycling" and the title of the horizontal axis is "Hours."

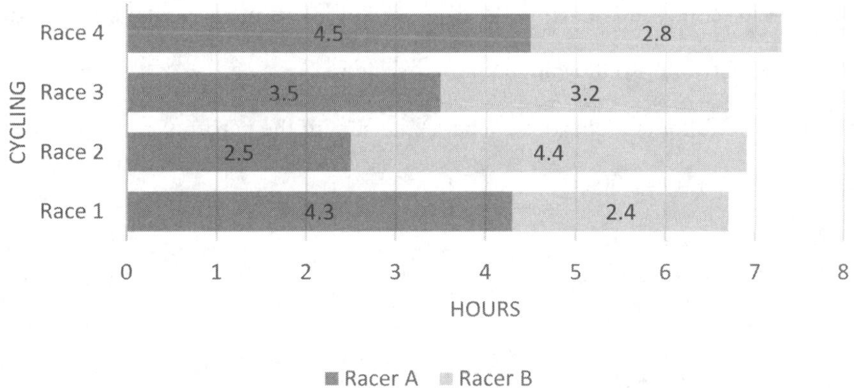

Example: Use the above bar chart to answer the following question: Which racer took longer to bike Race 2?

Solution: Racer A completed Race 2 in 2.5 hours. Racer B completed Race 2 in 4.4 hours. Therefore, Racer B took longer to bike Race 2.

Example: Use the above bar chart to answer the following question: If each racer's time was not stated in the bar, how would you know that Racer A and Racer B completed Race 3 in about the same time?

Solution: In the bar for Race 3, each racer's bar is about the same length. Therefore, their times are about the same.

Example: Use the above bar chart to answer the following question: How much longer did it take Racer A to bike Race 1 as compared to Racer B?

Solution: Racer A completed Race 1 in 4.3 hours. Racer B completed Race 1 in 2.4 hours. It took Racer A $4.3 - 2.4 = 1.9$ hours longer.

Pie Charts

A pie chart also represents categories. Instead of using bars, it uses pieces of a circle, or "pie," that are called sectors. The size of the sector depends on the size of the category. The sizes of the sectors are calculated mathematically so that they create a circle (i.e., a pie).

The pie chart below represents a financial portfolio, indicating the percent of the portfolio that is invested in stocks, bonds, CDs, and cash. Notice that 60% of the portfolio is invested in stocks and the stock sector makes up 60% of the pie chart.

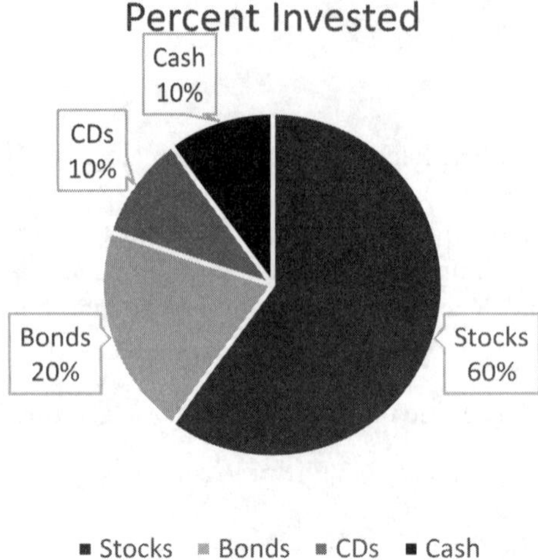

Example: Use the above pie chart to answer the following question: Which types of investments are equally invested?

Solution: CDs and Cash are both 10% of the portfolio, so they are equally invested.

Example: Use the above pie chart to answer the following question: What is the sum of the percents in the pie chart?

Solution: The sum is 100% because $10+10+20+60=100$.

Exercises

In Exercises 1–7, use the above stacked bar chart to answer the question.

1. How many new customers arrived at Store B during Week 3?
2. How many fewer new customers arrived at Store A during Week 2 than at Store B?
3. During which week did Store A and Store B receive about the same number of new customers?
4. Which store received more new customers in Week 4?

5. Week 1 has the longest bar overall. What is true about Week 1 for Store A? Store B?

6. How many new customers arrived at Store A over the four-week period?

7. How many new customers arrived at Store B over the four-week period?

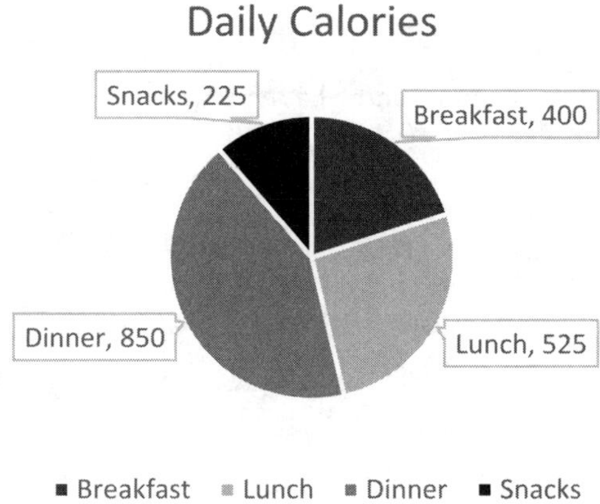

In Exercises 8–13, use the above pie chart to answer the question.

8. What is the total number of daily calories?

9. Breakfast and lunch together make up about what percent of the total daily calories?

10. Which meal has the largest number of calories?

11. What percent of the daily calories comes from snacks?

12. What percent of the daily calories comes from lunch and dinner?

13. Which meal has about 25% of the total daily calories?

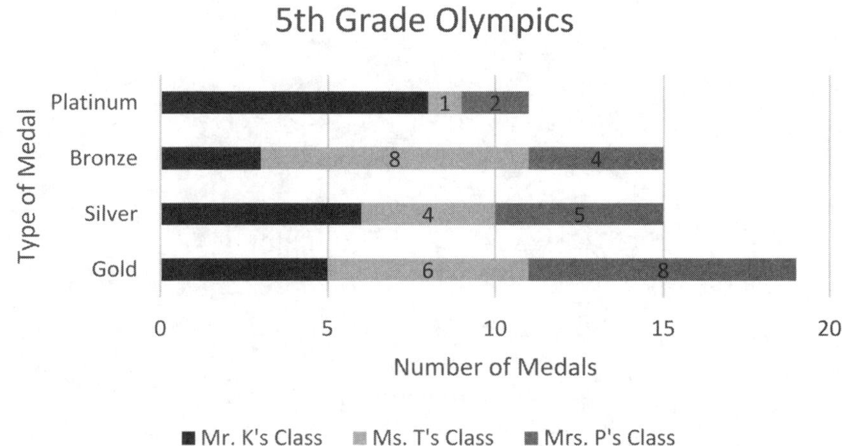

In Exercises 14–20, use the above stacked bar chart to answer the question.

14. How many students earned a gold medal overall?

15. How many more students earned a bronze medal in Ms. T's Class compared to Mr. K's Class?

16. Which medal was earned about the same amount in all three classes?

17. Which class earned the least number of overall medals?

18. Which class earned the greatest number of overall medals?

19. How many medals were earned overall?

20. Which class earned the least number of medals in any one type, and what type was it?

Chapter 1 Introduction to Statistics

Basic Mathematical Symbols and Greek Letters (Section 1.1)

Basic Mathematical Symbols

Mathematics involves symbols, letters, and numbers. This section will look at some basic mathematical symbols, modifications of letters, and Greek letters. You may find this to be a good reference section. Let's begin with symbols.

	Meaning	Example	Explanation
$<$	Less than	$x < 3$	All values less than but not including 3
$\le$	Less than or equal to	$x \le 3$	All values less than 3 and includes 3
$>$	Greater than	$x > 3$	All values greater than but not including 3
$\ge$	Greater than or equal to	$x \ge 3$	All values greater than and includes 3
$=$	Equal to	$x = 3$	Just 3
$\ne$	Not equal to	$x \ne 3$	Any value except 3 (includes less than and greater than)
$\approx$	Approximately	$x \approx 3$	A value close to 3, but not 3
$\pm$	Plus or minus	$x = \pm 3$	3 and -3

Letters are used to represent values that change. Letters are typed in italics. The English alphabet has 26 letters. This expands to 52 when distinguishing uppercase from lowercase. For example, N is the size of the population and n is the size of the sample. In this example, it is important that you distinguish between uppercase and lowercase.

In some situations, there are multiple occurrences of the same item. It would not make sense to use a different letter, so a subscript is utilized. For example, you let the cost of a textbook be denoted by the letter c. You purchase five textbooks and the costs of each textbook are denoted by the letters $c_1, c_2, c_3, c_4,$ and c_5. Can you see where this method is useful if you were keeping track of 500 textbooks?

In some subjects, such as statistics, letters are modified by adding a special character. The special character has a name, and it also has a special meaning. For example, $\bar{x}$ is pronounced "x-bar" and is the symbol for the sample mean, and $\hat{p}$ is pronounced "p-hat" and is the symbol for the sample proportion.

Greek Letters

Greek mathematicians have been credited with the discovery of many mathematical theorems and formulas. Thus, it should not be a surprise that some of the letters used in mathematical and statistical formulas are Greek letters. Here are some of the more popular ones.

	Name
α	Alpha, lowercase
β	Beta, lowercase
λ	Lambda, lowercase
μ	Mu, lowercase
ρ	Rho, lowercase
σ	Sigma, lowercase
Σ	Sigma, uppercase
χ	Chi, lowercase

Exercises

In Exercises 1–6, determine if the statement is true or false. Justify your answer.

1. If an uppercase letter is given, it is okay to use the lowercase form.
2. Given the set $x > 5$, 5 is in the set.
3. If the solution to an equation is $x = \pm 7$, the equation has only one solution.
4. If you round your answer, you should use the symbol $\approx$.
5. The only difference between $x < 10$ and $x \leq 10$ is that the second inequality includes 10 and the first inequality does not.
6. If you are using letters to keep track of the prices of 20 different trucks, then the best approach is to use 20 different letters of the alphabet.

In Exercises 7 and 8, write the modified letter and its name five times.

7. $\bar{x}$
8. $\hat{p}$

In Exercises 9–14, write the Greek letter and its name five times.

9. α
10. μ
11. σ
12. χ
13. Σ
14. ρ

In Exercises 15–20, write what each symbolic statement means.

15. $x < 1$
16. $x \geq 1$
17. $x \neq 1$
18. $x = 1$
19. $x \approx 1$
20. $x = \pm 1$

Chapter 1 Introduction to Statistics

Percents (Section 1.1)

Converting Percents to Fractions and Decimals

Percent means *per hundred* (a **cent**ury is one hundred years). So, 70% means 70 per hundred. Mathematically speaking, $70\% = \frac{70}{100}$. You can replace the % sign with a 100 in the denominator.

Example: Convert 100% to a fraction and simplify.

Solution

$$100\% = \frac{100}{100} = 1$$ 100% is equivalent to 1.

Because a fraction can be written as a decimal, $70\% = \frac{70}{100} = 0.70$. You can use a calculator to divide the fraction to get the decimal. You can also remove the percent sign and move the decimal point two places to the left.

Thus, a percent can be written with a percent sign, as a fraction, or as a decimal. Often the format is dependent on the context of the problem. It is important to be able to convert between the three forms.

Example: Convert 37.5% to a fraction and to a decimal. Do not simplify.

Solution

$$37.5\% = \frac{37.5}{100}$$ Convert to fraction.

$$= \frac{37.5}{100} \cdot \frac{10}{10}$$ Multiply by 1.

$$= \frac{375}{1000}$$ Decimal point is now removed in numerator.

$$= 0.375$$ Use a calculator or move decimal point two places left.

Converting Decimals and Fractions to Percents

In today's technological world, data will usually come to you in decimal form via a computer or a calculator. However, a percent is best represented with a percent sign. This process is just a reversal of what you did in the previous examples. If you have a decimal value, move the decimal point two places right and introduce a percent sign.

Example: Convert 0.26 to a percent.

Solution

$$0.26 = 26\%$$ Move decimal point two places right.

Tip: This is similar to the relationship between dollars and cents: $\$0.26 = 26¢$.

Example: Convert 1.4723 to a percent.

Solution

$$1.4723 = 147.23\%$$ Move decimal point two places right.

An alternative approach is to rewrite the decimal number as a fraction with a denominator of 100 and then write the number as a percent.

Example: Convert 1.4723 to a percent.

Solution

$$1.4723 = 1.4723 \cdot \frac{100}{100}$$ Multiply by 1 such that 100 is in the denominator.

$$= \frac{147.23}{100}$$ Simplify.

$$= 147.23\%$$ Convert to percent.

When the number is in fraction form, use a calculator to convert the number to decimal form. Then convert the number to a percent.

Example: Convert $\frac{23}{47}$ to a percent.

Solution

$$\frac{23}{47} \approx 0.48936$$ Convert to decimal and round.

$$= 48.936\%$$ Convert to percent.

Proportions

In statistics, the population proportion p is the proportion of the population (all people, places, or things that are being studied) that have a certain characteristic. For example, in the set of colors A = {pink, red, blue, purple, orange}, if p is the population proportion of colors that begin with the letter "p," then $p = \frac{2}{5}$ because there are 5 colors in the set and 2 of them begin with the letter "p."

Suppose a study is conducted that concludes that the population proportion p of U.S. babies born with a birth weight over 8 pounds is between 0.224 and 0.228. This is typical wording in a research report, but may be confusing to many people. The statement can be reworded by converting the decimal numbers to percents. The percentage of U.S. babies born with a birth weight over 8 pounds is between 22.4% and 22.8%.

Example: Rewrite the statement using percents, rounded to two decimal places.

The population proportion p of U.S. high school students participating in music or arts is between $\frac{1}{15}$ and $\frac{2}{15}$.

Solution

$\frac{1}{15} \approx 0.0667$ Convert to decimal and round.

$= 6.67\%$ Convert to percent.

$\frac{2}{15} \approx 0.1333$ Convert to decimal and round.

$= 13.33\%$ Convert to percent.

The percentage of U.S. high school students participating in music or arts is between 6.67% and 13.33% .

Exercises

In Exercises 1–6, convert to a fraction and simplify.

1. 74%

2. 18.5%

3. 40%

4. 85.4%

5. 12.25%

6. 0.5%

In Exercises 7–12, convert to a decimal.

7. 43%

8. 89.145%

9. 22.06%

10. 39.0021%

11. 6.34%

12. 0.57%

In Exercises 13–18, convert to a percent.

13. 0.61

14. $\frac{41}{100}$

15. $\frac{19}{53}$

16. 0.2974

17. 0.049

18. $\frac{2}{13}$

In Exercises 19 and 20, rewrite the statement using percents, rounded to two decimal places.

19. The population proportion p of members of the XYZ Fitness Center who eat yogurt at least five days a week is between $\frac{6}{11}$ and $\frac{8}{11}$.

20. The population proportion p of cats who like watching TV is between $\frac{15}{23}$ and $\frac{18}{23}$.

Chapter 1 Introduction to Statistics

Addition and Subtraction of Integers (Section 1.2)

What Is an Integer?

An **integer** is a **real number** that has no decimal part. Real numbers are your everyday numbers that you see on your calculator. Recall that a **set** is a collection of objects. The set of integers consists of the set of **natural numbers**, $\{1, 2, 3, ...\}$ (these are the **positive integers**), expanded to the set of **whole numbers**, $\{0, 1, 2, 3, ...\}$, (which introduces zero), and then expands to include the **negative integers**, which are the natural numbers multiplied by -1, $\{..., -3, -2, -1, 0, 1, 2, 3, ...\}$. The set of integers is infinite.

Eliminate Double Signs

One approach to adding and subtracting integers is to begin by eliminating any double signs. Rewrite $5+(+2)$ as $5+2$. Rewrite $3+(-1)$ as $3-1$. Rewrite $4-(+9)$ as $4-9$. Rewrite $7-(-8)$ as $7+8$. Notice that if the signs are the same, the result is addition. If the signs are different, the result is subtraction.

Example: Eliminate the double signs in the expression $5+(+2)-(+1)-(-6)+8+(-10)$.

Solution: If the signs are the same, then the result is addition. If the signs are different, then the result is subtraction.

$5+(+2)-(+1)-(-6)+8+(-10) = 5+2-1+6+8-10$ Eliminate the double signs.

Combine Like Terms

Once the double signs are eliminated, the expression consists of adding simple terms (the positive integers) and subtracting simple terms (the negative integers). **Terms** are values (in this case, integers) that are being added or subtracted. A common mathematical phrase is to "combine like terms." What this means is to add or subtract terms that are similar. In the above example, some of the terms are being added, so they are similar. If you combine them, you get $5+2+6+8=21$. Some of the terms are being subtracted, so they are similar. If you combine them, you get $-1-10=-11$. Notice that you just added the values and kept the sign of the terms—the sum of the positive terms is positive and the sum of the negative terms is negative. Your final expression is $21-11$.

Example: Combine like terms in the expression $14-3-6+11+4-9+7-20$.

Solution: $14-3-6+11+4-9+7-20 = 36-3-6-9-20$ Combine the positive terms.

$= 36-38$ Combine the negative terms.

Subtraction of Integers

Now that you have eliminated the double signs and combined like terms, the remaining expression is subtracting two integers (terms). When subtracting two integers, begin by determining if your answer will be positive or negative. If the positive integer is the largest, then your answer will be positive. If the negative integer is the largest, then your answer will be negative. In the above example, $36-38$, the negative integer is the largest so the answer will be negative. You can begin by writing a negative sign in your answer.

Example: Determine whether the answer is positive or negative $-42+17$.

Solution: Since the largest integer is negative, the answer is negative.

Since you have already determined that the answer is negative, to subtract the two integers, find the difference between the two integers 42 and 17, ignoring the signs. So, the final answer is -25

.

Example: Add or subtract the integers $16-5+(-11)-(+8)-(-12)$.

Solution: $16-5+(-11)-(+8)-(-12)=16-5-11-8+12$ Eliminate the double signs.

$=28-5-11-8$ Combine the positive terms.

$=28-24$ Combine the negative terms.

Since the largest integer is positive, the answer will be positive.

$=4$ Find the difference between the two integers.

Exercises

In Exercises 1–4, eliminate the double signs in the expression.

1. $-15-(+4)+(+16)+(-9)$

2. $28+(-19)-(-17)-(-1)+(-8)-(+14)$

3. $3-(-2)+(-7)-(+6)+9-10$

4. $-36+(-15)+(-21)-60-(-24)-(+36)+(+100)$

In Exercises 5–10, combine like terms (combine the positive terms and combine the negative terms).

5. $13-8-10+4-5+7$

6. $-17-22+10-25+5+16-1+14$

7. $29+31-34-16-2+36+1-0+19$

8. $56-62-91+45+87-75-72-83$

9. $-100-105+124-106+110+125-140$

10. $-300+500+800-900-600+200-300-400+500$

In Exercises 11–20, add or subtract the integers.

11. $15+(-6)-(+10)-18$

12. $-21-17+(-9)+10-(-2)$

13. $-30-(-11)+(+5)-17-(+10)+8+(-29)$

14. $1-51+(-19)-40-(-22)-(+20)+(+31)+24-(-15)$

15. $-45+(+62)+(-74)-(+61)-(-58)-50$

16. $-88-(-88)+(-56)-(-64)+(+48)-(-92)+100$

17. $60+(-55)-(-35)+(+70)-(+85)+(-90)$

18. $-112-(-118)+(-86)+(+127)-(+96)-(-121)-125$

19. $400-(+300)+(-900)-(-800)-(+400)-(-700)+300$

20. $-512+(-256)-(-36)+(+129)-(-456)-(+518)+58-912$

Chapter 2 Descriptive Statistics

Multiplication and Division of Integers (Section 2.1)

Multiplying Integers

The result of multiplication is called a **product.** The product of two integers can be written in a variety of ways. For example,

$$(2)(5), \qquad 2(5), \qquad (2)5, \qquad 2\times5, \qquad 2\cdot5$$

all represent the product of "2 times 5," which equals 10. The two integers, 2 and 5, are called **factors.** Factors are values (in this case, integers) that are being multiplied or divided.

- When multiplying two negative integers, the product is positive. For example, $(-1)(-7)=7$.
- When multiplying one positive integer and one negative integer, the product is negative. For example, $-9\cdot6=-54$.
- When multiplying an integer by zero, the product is zero. For example, $3\times0=0$.

To find the product of more than two integers, begin by finding the product of any two of the integers, resulting in a new multiplication problem. Continue this process until the final product is found. For example, to find the product of $(5)(7)(4)$, an efficient first product would be to find $(5)(4)=20$ because 20 is a nice number to multiply. This leaves a new multiplication problem of $(20)(7)$. The final product is 140.

Example: Find the product $(3)(-6)(-3)(-4)$.

Solution: $(3)(-6)(-3)(-4) = (-9)(-6)(-4)$ $\quad (3)(-3)=-9$

$= (54)(-4)$ $\quad (-9)(-6)=54$

$= -216$ $\quad$ Final product

Dividing Integers

The result of division is called a **quotient.** The quotient of two integers can be written in a few ways. For example,

$$18\div6, \qquad 18/6, \qquad \frac{18}{6}$$

all represent the quotient of "18 divided by 6," which equals 3. Since 6 is divided into 18, it is the **divisor.** The 18 is being divided, so it is the **dividend.** The dividend, 18, is in the **numerator** of the fraction $\frac{18}{6}$ and the divisor, 6, is in the **denominator.**

Division can be written in terms of multiplication. $\frac{18}{6}=3$ because $6\cdot3=18$. Notice that 18 is the dividend and the product. The rules of multiplication are slightly modified for division.

- When dividing two negative integers, the quotient is positive. For example, $\frac{-14}{-7} = 2$ because $(-7)(2) = -14$.

- When dividing one positive integer and one negative integer, the quotient is negative. For example, $\frac{-20}{2} = -10$ because $2 \cdot -10 = -20$.

- When dividing zero by an integer, the quotient is zero. For example, $\frac{0}{11} = 0$ because $11 \times 0 = 0$.

- When dividing an integer by zero, the quotient is **undefined.** It is undefined because this division cannot be written in terms of multiplication. For example, $\frac{4}{0}$ is undefined because 0 times any integer equals 0, not 4.

Exercises

In Exercises 1–10, find the product.

1. $-16 \cdot 2$

2. 0×5

3. $(-7)(-9)$

4. $6 \cdot 7 \cdot 5$

5. $2 \times 9 \times 4$

6. $5 \cdot 5 \cdot 6 \cdot 6$

7. $(-2)(-4)(-11)$

8. $(3)(-5)(8)$

9. $(-10)(4)(-3)(-6)$

10. $(12)(-15)(-25)(14)(-42)$

In Exercises 11–19, find the quotient.

11. $21 \div (-3)$

12. $0/5$

13. $\frac{-24}{-3}$

14. $12 \div 0$

15. $\frac{36}{-18}$

16. $-64/16$

17. $\frac{0}{-19}$

18. $(-100) \div (-5)$

19. $\frac{-25}{0}$

20. Explain why $\frac{9}{0} = 0$ is not a true statement.

Chapter 2 Descriptive Statistics

Order of Operations (Section 2.1)

Add, Subtract, Multiply, and Divide—Who Goes First?

Multiply and divide go first and they share first position. Whichever one arrives first takes the lead. Once they have finished their tasks, then add and subtract go to work, sharing a position. Even though there are four operations (add, subtract, multiply, and divide), there are only two shifts (the first shift is the multiply–divide shift, and the second shift is the add–subtract shift).

For example, if you have $4-12\div2\times3+1$, then you divide first (you evaluate expressions from left to right and division occurs before multiplication). The expression is then $4-6\times3+1$. You are still in the multiply–divide shift, so multiply next. The expression is now $4-18+1$. Multiply and divide have now finished their tasks, so you are in the add–subtract shift. This is where you use the skills that you learned in Addition and Subtraction of Integers. The final answer is -13.

Example: Evaluate $10+18/6\times3-2$.

Solution: $10+18/6\times3-2=10+3\times3-2$	Division occurs first from the left.
$=10+9-2$	Multiplication occurs next from the left.
$=17$	Add or subtract.

Grouping

The priority of the operations can be overridden with grouping. If the above expression $4-12\div2\times3+1$ is rewritten as $(4-12)\div2\times3+1$, then you would not divide first. The operation within the grouping, subtraction, now has priority. The expression is then $-8\div2\times3+1$. Now you are in the multiply–divide shift. First from the left is division. The expression is then $-4\times3+1$. To finish the multiply–divide shift, you multiply to get $-12+1$. The add–subtract shift gives you a final answer of -11.

Notice that the answers are different with and without grouping. Without grouping, the answer is -13, and with grouping, the answer is -11. Thus, order is important. Order is not optional.

This order is universal and is programmed into calculators and computers. The order is:

1. Grouping
2. Multiplication–division
3. Addition–subtraction

The above example used parentheses, $(\ )$, for grouping. However, brackets $[\ ]$, braces $\{\ \}$, and other symbols can also be used. The grouping expression, $(4-12)\div2\times3+1$, can be written as $[4-12]\div2\times3+1$ or $\{4-12\}\div2\times3+1$.

The most common way to rewrite the expression $(4-12)\div2\times3+1$ is to use fractions, $\frac{4-12}{2}\times3+1$. Notice that the numerator is used for grouping and the horizontal fraction bar is used for division. Grouping has priority over division, so $4-12$ is the first operation performed. Notice that the parentheses were removed. They are implied, but they are not needed when written in fraction form. If you were to type this into a calculator, you would need to put the parentheses around the numerator, because the calculator does not "imply."

Example: Evaluate $\frac{10+8}{(2)(3)}-5$.

Solution: There are parentheses in the denominator, but they are multiplication symbols, not grouping symbols. Because there are operations in both the numerator and the denominator, there is grouping in both numerator and denominator.

$\frac{10+8}{(2)(3)}-5=\frac{18}{6}-5$ Perform the operations in the groupings first.

$=3-5$ Division has priority over subtraction.

$=-2$ Subtract.

Exercises

In Exercises 1–20, evaluate the expression.

1. $4\times6+10\div2+9$
2. $16\div8-7\times2+6$
3. $-12+4\div(-2)\cdot5+(-7)$
4. $20-(-4)\cdot12\div6-(-10)$
5. $15+30\div(-3)\div(-5)\cdot(-2)$
6. $38-6\cdot(-15)\div(3)\div2$
7. $4+(5-7)(-3)\div2+10$
8. $7-(10+4)\cdot3\div7-15$
9. $36\div[(-12)\div2]+9\cdot3-5$
10. $25((-20)\div(-5))-8\cdot4-10$
11. $\frac{(2)(5)+(3)(7)+(4)(10)+(5)(13)}{1+2+5}$
12. $\frac{(3)(-2)+(-6)(-1)+(-7)(7)+(9)(-8)}{21-10}$
13. $\frac{(6)(4)+(6)(8)+(6)(-5)+(6)(-13)+(6)(11)+(6)(14)}{3+9+7}$
14. $\frac{19-1}{6}\cdot12-28/7+3$
15. $\frac{16+5}{7}\cdot5+35\div7-2$

16. $\dfrac{5\cdot1+5\cdot2+5\cdot3+5\cdot4+5\cdot5+5\cdot6}{1\cdot3\cdot5\cdot7}$

17. $12\left(\dfrac{62-45}{2\cdot3}\right)-24/8+17$

18. $\left(\dfrac{4\cdot3+5\cdot7+6\cdot10+7\cdot13}{2}\right)\cdot0$

19. $8\left[\dfrac{48\div4}{4\cdot4}\right]\div6$

20. $\dfrac{7\cdot6\cdot5\cdot4\cdot3\cdot2\cdot1}{(3\cdot2\cdot1)\cdot(4\cdot3\cdot2\cdot1)}$

Chapter 2 Descriptive Statistics

Operations with Fractions and Decimals (Section 2.1)

Reducing Fractions

When an answer is in fraction form, you are expected to reduce it if possible. A fraction in reduced form does not have any common factors between the numerator and the denominator. For example, $\frac{10}{6}$ is not in reduced form because both 10 and 6 have a common factor of 2 $\left(\frac{10}{6}=\frac{2\cdot 5}{2\cdot 3}\right)$. The common factor can be divided out, leaving a fraction of $\frac{5}{3}$. Both 5 and 3 are prime numbers (divisible only by themselves and 1), so you now have a fraction in reduced form. Many mathematical software programs will mark your fractional answer incorrect if it is not in reduced form.

Example: Reduce $\frac{24}{15}$.

Solution: $\frac{24}{15}=\frac{3\cdot 8}{3\cdot 5}$ Factor out the common factor of 3.

$=\frac{\not{3}\cdot 8}{\not{3}\cdot 5}$ Divide out the common factor.

$=\frac{8}{5}$ Simplify.

Since 5 is a prime number and 8 is not divisible by 5, the fraction is in reduced form.

Multiply Fractions

When multiplying two fractions, multiply the numerators and multiply the denominators, then reduce if possible. For example, when multiplying $\frac{2}{5}$ by $\frac{7}{4}$, begin by multiplying 2 by 7 and then 5 by 4. This gives you a product of $\frac{14}{20}$. This fraction can then be written as $\frac{2\cdot 7}{2\cdot 10}$. The common factor of 2 can be divided out, leaving a fraction of $\frac{7}{10}$. The numerator, 7, is a prime number and is not a factor of the denominator. Thus, the fraction is in reduced form.

Example: Evaluate $\frac{3}{8}\cdot\frac{10}{9}$.

Solution: $\frac{3}{8}\cdot\frac{10}{9}=\frac{30}{72}$ Multiply numerators and denominators.

$=\frac{6\cdot 5}{6\cdot 12}$ Factor out common factor.

$= \frac{\not{6} \cdot 5}{\not{6} \cdot 12}$ Divide out common factor.

$= \frac{5}{12}$ Simplify; this is in reduced form.

Divide Fractions

When dividing two fractions, begin by rewriting it as a multiplication problem. For example, when dividing $\frac{7}{12}$ by $\frac{3}{4}$, rewrite as multiplying $\frac{7}{12}$ by $\frac{4}{3}$. Notice that the first fraction did not change. The second fraction changed to its *reciprocal*. Now, it is a multiplication problem.

Example: Evaluate $\frac{9}{5} \div \frac{3}{20}$.

Solution: $\frac{9}{5} \div \frac{3}{20} = \frac{9}{5} \cdot \frac{20}{3}$ Multiply by the reciprocal.

$= \frac{(9)(20)}{(5)(3)}$ Multiply numerators and denominators.

When the products get larger, it is helpful to leave in factored form.

$= \frac{(3)(3)(5)(4)}{(5)(3)}$ Factor out common factors.

$= \frac{\not{(3)}(3)\not{(5)}(4)}{\not{(5)}\not{(3)}}$ Divide out common factors.

$= \frac{(3)(4)}{1}$ Simplify.

$= \frac{12}{1}$ Multiply.

$= 12$ Divide.

Add and Subtract Fractions

When adding or subtracting two fractions, they must have the same denominator. For example, when adding $\frac{1}{6}$ and $\frac{2}{3}$, a common denominator is 6. To rewrite $\frac{2}{3}$, you must be careful not to change its value. Multiplying by 1 does not change its value: $\frac{2}{3} \cdot 1 = \frac{2}{3} \cdot \frac{2}{2} = \frac{4}{6}$. The addition problem now becomes $\frac{1}{6} + \frac{4}{6} = \frac{1+4}{6} = \frac{5}{6}$. Notice that the denominator stayed the same, and you only added the numerators.

Example: Evaluate $\frac{2}{9} - \frac{5}{6}$.

Solution: The fractions do not have the same denominator. 18 is divisible by both 9 and 6 .

$$\frac{2}{9} - \frac{5}{6} = \frac{2}{9} \cdot \frac{2}{2} - \frac{5}{6} \cdot \frac{3}{3}$$ Multiply by 1 to get common denominator.

$$= \frac{4}{18} - \frac{15}{18}$$ Multiply.

$$= \frac{4 - 15}{18}$$ Subtract

$$= -\frac{11}{18}$$ Simplify; this is in reduced form.

Fractions with Decimals

Fractions may contain decimal numbers rather than whole numbers. The operations of multiplication, division, addition, and subtraction are the same. Reducing fractions may be different, however. Usually a calculator or long division is used to reduce the fraction.

Example: Evaluate $\frac{1.24 + 2.25 + 2.27 + 2.42 + 2.31}{5}$ and simplify. Round to two decimal places.

Solution: $$\frac{1.24 + 2.25 + 2.27 + 2.42 + 2.31}{5} = \frac{10.49}{5}$$ Add the numerator.

$$= 2.098$$ Divide.

$$= 2.10$$ Round to two decimal places.

Example: Evaluate $\frac{1.1}{3} \cdot \frac{10.5}{8}$ and simplify. Round to two decimal places.

Solution: $$\frac{1.1}{3} \cdot \frac{10.5}{8} = \frac{11.55}{24}$$ Multiply numerators and denominators.

$$= 0.48125$$ Divide.

$$= 0.48$$ Round to two decimal places.

Exercises

In Exercises 1–4, reduce the fraction.

1. $\frac{18}{12}$

2. $\frac{27}{45}$

3. $\frac{32}{88}$

4. $\frac{210}{150}$

In Exercises 5–8, multiply the fractions.

5. $\frac{8}{9}\cdot\frac{6}{20}$

6. $\frac{10}{3}\cdot\frac{21}{25}$

7. $\frac{7}{6}\cdot\frac{42}{49}$

8. $\frac{18}{11}\cdot\frac{44}{15}$

In Exercises 9–12, divide the fractions.

9. $\frac{7}{12}\div\frac{21}{40}$

10. $\frac{2}{35}\div\frac{18}{45}$

11. $\frac{10}{39}\div\frac{25}{12}$

12. $\frac{64}{33}\div\frac{72}{121}$

In Exercises 13–16, add or subtract the fractions.

13. $\frac{2}{5}-\frac{3}{10}$

14. $\frac{3}{4}+\frac{5}{6}$

15. $\frac{7}{8}+\frac{3}{20}$

16. $\frac{10}{21}-\frac{2}{9}$

In Exercises 17–20, perform the operation and simplify. Round to two decimal places.

17. $\frac{2.1}{3}+\frac{1}{6}$

18. $\left(\frac{17.5}{20}\right)\left(\frac{6.5}{12}\right)$

19. $\frac{5}{6}-\frac{2.3}{10}$

20. $\frac{9.5}{12}\div\frac{5}{6}$

Chapter 2 Descriptive Statistics

Significant Digits and Rounding (Section 2.1)

When you report measurements or calculations, significant digits and rounding guide how you represent the numbers. Some fields of research have specific guidelines that are used in their field. In this section, you will learn some general guidelines.

Basic Concept of Significant Digits

The basic concept is that the significant digits are the number of digits beginning with the first nonzero digit and counting to the right. For example, 475 has three significant digits. The first digit, 4 , is a nonzero digit and there are two more digits to the right. Thus, there are three significant digits.

Example: Determine the number of significant digits: 0.05609 .

Solution: The first digit is zero, which is not a significant first digit. The first nonzero digit is 5 , and there are three more digits to the right. Therefore, there are four significant digits.

Exceptions: Significant Digits and Rounding

Significant digits are also known as accurate digits (as accurate as measurements can get). According to the above basic concept, 400 has three significant digits. If this was the reported balance of your checking account, then if you asked for four $100 bills, you would be able to withdraw that amount from your account and your balance would be $0. In this scenario, 400 does have three significant digits.

What if the bank estimated your account balance? You have $389 in your account. The bank rounded to the hundreds. In the rounding to the nearest hundred, looking to the digit to the right of the hundreds place, you see that the 8 is 5 or larger, so the 3 is rounded up to a 4 . In this scenario, 400 only has one significant digit because it is only accurate to the hundreds place (and not really accurate, because you do not actually have $400).

If your bank rounded to the tens place and you have $403 in your account, then 400 has two significant digits (3 is less than 5 , so no rounding occurs). However, if you have $405 in your account, the bank would have estimated your account balance at $410 (5 is greater than or equal to 5 , so round up), which still has two significant digits.

Therefore, 400 could have one, two, or three significant digits depending on the rounding. For large numbers, it is difficult to know if the zeroes are only needed as placeholders or if they are significant digits.

However, 400.0 has four significant digits. The zero to the right of the decimal point is not there just as a placeholder. It was specifically put there. The zero occurred due to rounding, and therefore, is significant. An example of a number prior to rounding is 400.03954 .

Example: Determine the number of significant digits: 1600

Solution: The 1 and 6 are significant digits. It is difficult to know if the two zeros are placeholders or significant digits. Therefore, there could be two, three, or four significant digits.

Example: Determine the number of significant digits: 56.20

Solution: Since the zero at the end is to the right of the decimal point, it occurred due to rounding. Therefore, it is significant. There are four significant digits.

Example: Round to three significant digits: 0.018453.

Solution: The first significant digit is 1. Rounding will occur after the 4, 0.0184|53. The next digit is 5, so the 4 will round up to a 5. The number rounded to three significant digits is 0.0185.

Example: Round to three significant digits: 78,325.

Solution: The first significant digit is 7. Rounding will occur after the 3, 78,3|25. The next digit is 2, so the 3 will not round up. The number rounded to three significant digits is 78,300.

Exercises

In Exercises 1–10, determine the number of significant digits.

1. 2067

2. 0.0023

3. 60

4. 25,000

5. 599.50

6. 0.097800

7. 750,139

8. 40.1257

9. 1,000.00

10. 0.0000006789

In Exercises 11–20, round to three significant digits.

11. 0.2876651

12. 5209

13. 12.348

14. 0.0094157

15. 68,037

16. 54,076

17. 3.14159

18. 0.0779612

19. 154.65

20. 0.012345

Chapter 2 Descriptive Statistics

The Number Line and Ordering Numbers (Section 2.2)

The Number Line

The number line is one dimensional. The one shown below goes from left to right, which is how you will usually see it. Sometimes you will see it drawn vertically.

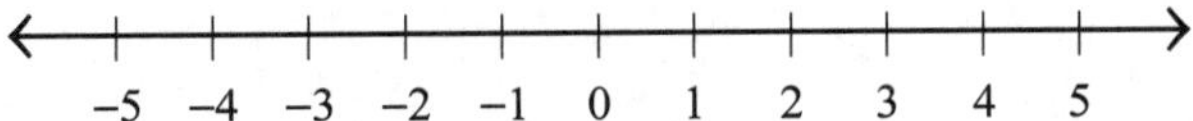

The arrows on the ends indicate that the line never ends. The arrows are important; they are part of the number line. The right side extends to positive infinity, $+\infty$, and the left side extends to negative infinity, $-\infty$.

The number zero, 0, separates the negative numbers from the positive numbers. Since the left side extends to $-\infty$, the negative numbers are to the left of 0. Since the right side extends to $+\infty$, the positive numbers are to the right of 0. If the number line is drawn vertically, then the negative numbers are below 0 and the positive numbers are above 0.

This is the real number line. The above number line is marked with whole numbers, but there are infinitely many real numbers in between each of those whole numbers. Every real number corresponds to exactly one point on the real number line, and there are infinitely many real numbers.

The number line above is showing you the numbers around -5 to 5. It is showing you just a small portion of the number line. The number line below is showing you a different portion of the same number line. It is showing you the numbers around 100 to 110. You zoom in to the portion that you need.

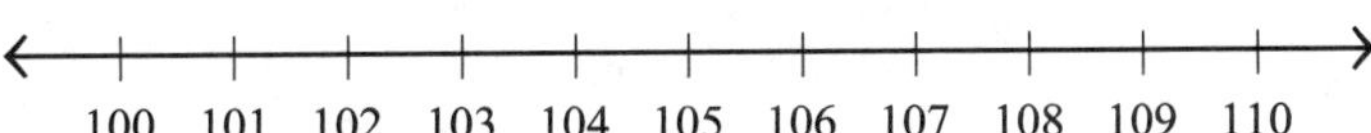

The number lines shown above are counting by 1. Depending on the spread of the numbers, it is sometimes necessary to count by a larger or smaller number. The first number line below is counting by 10. The second number line is counting by 0.5.

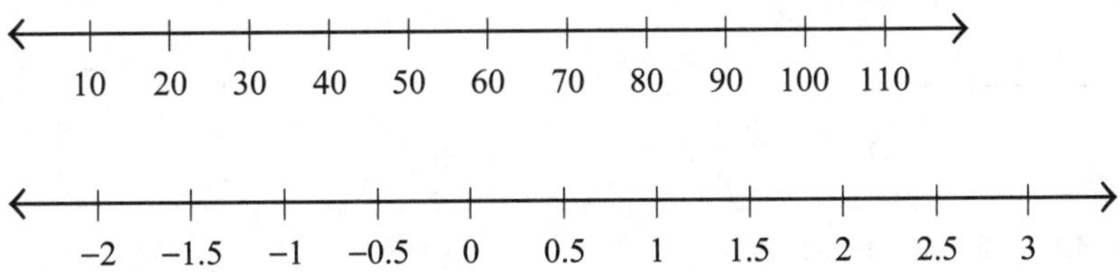

Ordering Numbers from Least to Greatest

When collecting numerical data in the real world, the numbers are often placed in a list in the order in which they are collected. When analyzing these data, it is sometimes important for the numbers to be in order from least to greatest. Least to greatest could also be expressed as smallest to largest, but this wording can be confusing. One might think that 2 is smaller than -10, but it is not. On the number line, -10 is to the left of 2. Therefore, $-10 < 2$, so -10 is the least and 2 is the greatest. The number line can help you visualize the least and the greatest numbers.

Example: Order the numbers from least to greatest: $10, 3, -2, 0, 5, -4, -5$.

Solution: The list contains both positive and negative numbers. The negative numbers are to the left of 0 on the number line, so they are the least numbers. The positive numbers are the greatest numbers. One approach is to plot the numbers as points on a number line.

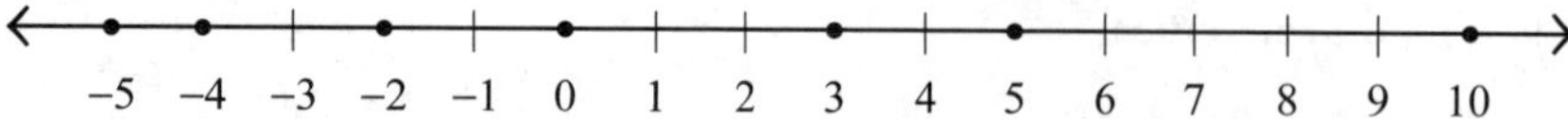

Moving from left to right, the order from least to greatest is $-5, -4, -2, 0, 3, 5, 10$. Notice that the magnitude of the negative numbers (the value without the sign; the magnitude of -5 is 5) gets larger as you move to the left on the number line, creating least numbers. The magnitude of the positive numbers gets larger as you move to the right on the number line, creating greatest numbers.

Example: Order the numbers from least to greatest: $-8, -1, 13, 10, 15, -9, -16, 7$.

Solution: The list contains both positive and negative numbers. The negative number with the largest magnitude is -16, so that is the least number. The positive number with the largest magnitude is 15, so that is the greatest number. So far, your list from least to greatest looks like $-16, \ldots, 15$ and your original list has the remaining numbers: $-8, -1, 13, 10, \cancel{15}, -9, \cancel{-16}, 7$.

From the remaining numbers, the negative number with the largest magnitude is -9, and the positive number with the largest magnitude is 13. Your list from least to greatest now looks like $-16, -9, \ldots, 13, 15$ and your original list has the remaining numbers:
$-8, -1, \cancel{13}, 10, \cancel{15}, \cancel{-9}, \cancel{-16}, 7$.

Continuing on, the next least number is -8, and the next greatest number 10. Your list from least to greatest now looks like $-16, -9, -8, \ldots, 10, 13, 15$ and your original list has the remaining numbers $\cancel{-8}, -1, \cancel{13}, \cancel{10}, \cancel{15}, \cancel{-9}, \cancel{-16}, 7$.

It is easy to see where the two remaining numbers fit in to the list. The numbers ordered from least to greatest are $-16, -9, -8, -1, 7, 10, 13, 15$.

Exercises

1. Label the number line around -3 to 3, counting by 1.

2. Label the number line around 30 to 38, counting by 1.

3. Label the number line around 62 to 72, counting by 1.

4. Label the number line around -6 to 0, counting by 1.

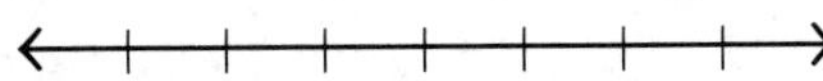

5. Label the number line around 0 to 16, counting by 2.

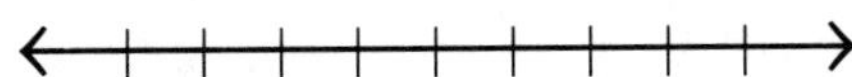

6. Label the number line around -25 to 25, counting by 5.

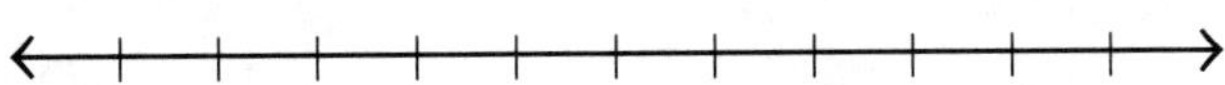

7. Label the number line around 0 to 4, counting by 0.5.

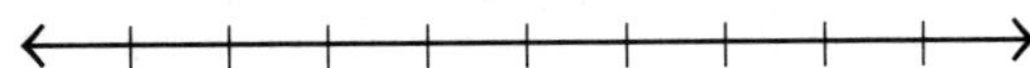

8. Label the number line around 10 to 10.5, counting by 0.1.

9. Label the number line around 0 to 1, counting by 0.25.

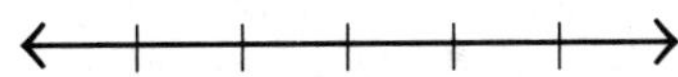

10. Label the number line around -1 to 0, counting by $\frac{1}{4}$.

In Exercises 11–20, order the numbers from least to greatest.

11. $12, -6, 7, 9, -3, -14, 11$

12. $-16, -14, 8, 0, -5, -10, 14, 20$

13. $24, -52, 18, 40, -35, -26, -61, 29, 50$

14. $104, -96, 75, 39, -73, -114, 121$

15. $-350, -296, -375, -319, -273, -314, -321$

16. $1.5, 2.75, -3.5, -1, 3.25, -1.25, -1.5, 2.25$

17. $-6.3, 5.8, -7.9, 7.4, -8.1, -8.5, 6.5, 7.5$

18. $-92, -87.58, 79, -54.10, -81, 85.23, -65.15, 75.18$

19. $-\frac{5}{12}, \frac{1}{12}, -\frac{7}{12}, -\frac{1}{12}, \frac{5}{12}, \frac{7}{12}$

20. $\frac{5}{12}, \frac{1}{12}, \frac{1}{4}, \frac{1}{2}, \frac{3}{4}, \frac{7}{12}, \frac{5}{6}$

Chapter 2 Descriptive Statistics

The *xy*-Plane and Point Plotting(Section 2.1)

The *xy*-Plane

The *xy*-plane consists of a horizontal number line (the ***x*-axis**) and a vertical number line (the ***y*-axis**) that intersect at 0 on each of the number lines, as you can see in the *xy*-plane below. The positive *x*-values are on the right and the negative *x*-values are on the left. The positive *y*-values are on the top and the negative *y*-values are on the bottom.

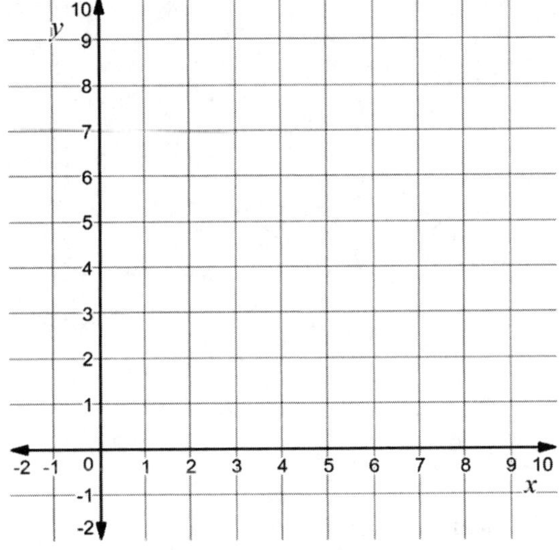

Unless you are doing some type of special graphing, the *x*-axis is always the horizontal axis and the *y*-axis is always the vertical axis. If you choose to be different, your graphs will not only look different, but they may also not have the mathematical characteristics that they should have.

The horizontal axis (*x*-axis) is the **independent axis** and the vertical axis (*y*-axis) is the **dependent axis**. In most instances, the value of *y* depends on the chosen value of *x*. The value of *x* is independent. In algebraic equations, you choose values for *x*. In research, *x*-values usually represent time (days, months, etc.), which you cannot completely control.

Point Plotting

The axes are number lines so they extend indefinitely, which is indicated by the arrows. This extension creates the *xy*-plane. You can think of the *xy*-plane as being a piece of paper that extends forever. The *xy*-plane has four **quadrants** (regions), separated by the axes.

The *xy*-plane consists of points, that are referenced according to their relationship with the *x*- and *y*-axes. Points are designated as (x, y), where *x* is the ***x*-coordinate** and *y* is the ***y*-coordinate**. Points are not designated as x, y, which would be a list of two values or variables. When designating a point, the parentheses, (), are not optional. The point where the *x*- and *y*-axes intersect is called the **origin** and its coordinates are $(0, 0)$. The point $(2, 4)$ is plotted at the intersection of the vertical line crossing the *x*-axis at 2 (the *x*-coordinate) and the horizontal line crossing the *y*-axis at 4 (the *y*-coordinate).

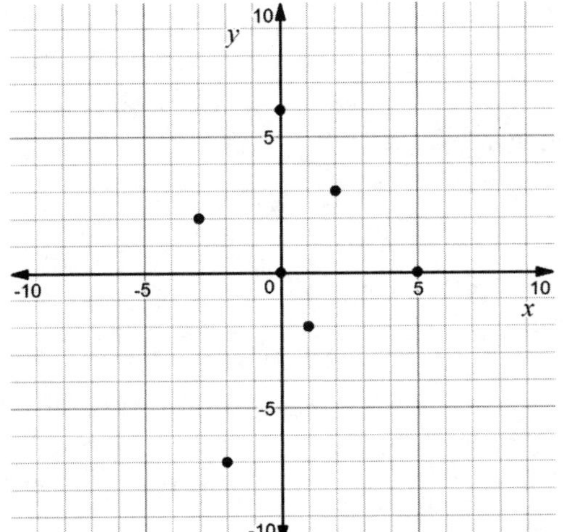

Label	x	y
A	0	0
B	5	0
C	0	6
D	2	3
E	−3	2
F	1	−2
G	−2	−7

Example: There are seven points plotted in the *xy*-plane at the left. Label them according to the following table.

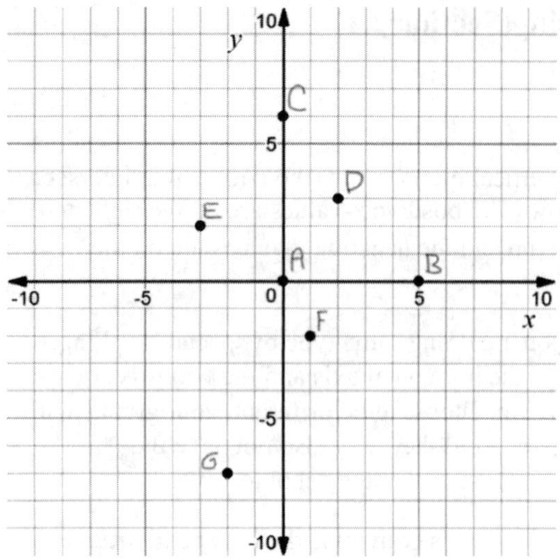

Solution: The seven points are labeled in the *xy*-plane at the left.

Exercises

1. Which axis is the horizontal axis?
2. Which axis is the dependent axis?
3. What is the name of the point where the *x*- and *y*-axes intersect? What are its coordinates?
4. The axes divide the *xy*-plane into four regions called what?
5. Given the point $(-2, 6)$, what is the value of the *y*-coordinate?
6. Given the point $(18, -3)$, what is the value of the *x*-coordinate?
7. When asked to write the coordinates of a point, a student wrote: $9, 1$. State why the student's answer is incorrect and write the answer correctly.
8. On the *y*-axis, are the negative numbers on the top or the bottom?
9. On the *x*-axis, are the positive numbers on the left or the right?
10. True or false: The *xy*-plane is a collection of points.
11. True or false: When designating a point, the parentheses, (), are optional.

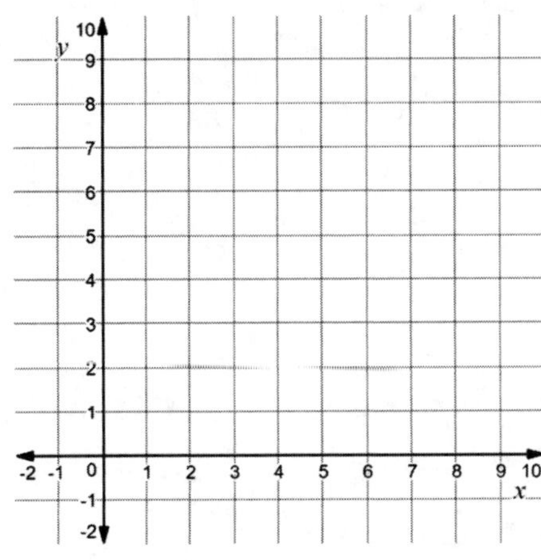

In Exercises 12–20, plot and label the point on the *xy*-plane at the left.

12. $A(0, 0)$

13. $B(1, 4)$

14. $C(2, 5)$

15. $D(3, 3)$

16. $E(4, 5)$

17. $F(5, 7)$

18. $G(6, 9)$

19. $H(7, 6)$

20. $I(8, 0)$

Chapter 2 Descriptive Statistics

Summation (Section 2.1)

Simple Summation

If you wanted to write the summation of the first five positive integers, you would write $1+2+3+4+5$. It would be difficult to write the summation of the first 500 positive integers this way! Since there are many instances of summing up large amounts of numbers, mathematics has a shortcut for writing summation. The uppercase Greek letter **sigma,** Σ, is the mathematical symbol used for summation.

Here are the first five positive integers, presented as values of *x*.

x	1	2	3	4	5

You can represent the summation of these first five positive integers by writing Σx. When evaluated, $\Sigma x = 1+2+3+4+5 = 15$, the sum of the first five positive integers. If the first eight positive integers were presented as values of *x*, then Σx would sum up the first eight positive integers. Basically, Σx sums up the *x*-values.

Example: Given the values of *y*, evaluate Σy.

y	–5	0	4	9	17

Solution: The summation of *y* is the sum of the *y*-values.

$$\Sigma y = -5+0+4+9+17 = 25$$

Summation of an Expression

Often, the summation is more involved than just adding up some numbers. The table contains data presented as values of *f* and as values of *m*. Each column of data came from the same data source. The sum of the *f*-values is $\Sigma f = 2+4+5+3+6+2 = 22$ and the sum of the *m*-values is $\Sigma m = 12+11+9+8+10+12 = 62$.

f	2	4	5	3	6	2
m	12	11	9	8	10	12

You are interested in the product of the *f*- and the *m*-values for each data source. You begin by creating a new row in the table, labeling the row as $f \cdot m$, and then multiplying the values in each column ($2 \cdot 12 = 24$, $4 \cdot 11 = 44$, etc.). You can now evaluate the sum of the $f \cdot m$ row, $\Sigma(f \cdot m) = 24+44+45+24+60+24 = 221$.

f	2	4	5	3	6	2
m	12	11	9	8	10	12
$f \cdot m$	24	44	45	24	60	24

Example: Given the values of *x*, evaluate $\Sigma(x-2.1)$.

x	8	12	13	9	7	4

Solution: First, 2.1 must be subtracted from each *x*-value. A new row is created, the row is labeled as $x-2.1$, and 2.1 is subtracted from each *x*-value. Then $\Sigma(x-2.1)=5.9+9.9+10.9+6.9+4.9+1.9=40.4$.

x	8	12	13	9	7	4
$x-2.1$	5.9	9.9	10.9	6.9	4.9	1.9

Exercises

In Exercises 1–8, evaluate Σx, given the value of *x*.

1.

x	5	13	6	9	0	4

2.

x	9	4	8	16	10	3	1

3.

x	12	–7	–10	9	16	5	8	2

4.

x	–17	–9	14	8	–15	14	12	–11

5.

x	3.2	1.25	5.5	4.7	3.1	2.75

6.

x	0.6	–1.8	1.2	–0.3	–1.1	0.9

7.

x	23.4	–7.3	–14.5	9.6	10.5	–22.1	–15.6	13.4

8.

x	–30.5	89.4	74.89	–61.56	–94.31	–101.52	112.45	96.78	–99.98	–67.45

9. Given the values of *y*, evaluate $\Sigma(5y)$.

y	7	10	3	6	12	11	1	9	2

10. Given the values of *y*, evaluate $\Sigma\left(\frac{1}{2}y\right)$.

y	10	20	30	40	50	60	70	80	2

11. Given the values of *y*, evaluate $\Sigma(-3y)$.

y	12	6	–2	8	–12	–9	4	–5	7

12. Given the values of *y*, evaluate $\Sigma(-0.1y)$.

y	–65	15	40	85	–75	–60	55	35	–45

13. Given the values of *w* and *x*, evaluate $\Sigma(w \cdot x)$.

w	3	6	4	9	8
x	1	3	5	7	9

14. Given the values of *w* and *x*, evaluate $\Sigma(w \cdot x)$.

w	0.2	0.1	0.4	0.1	0.2
x	80	75	90	60	95

15. Given the values of *w* and *x*, evaluate $\Sigma(w \cdot x)$.

w	0.4	0.25	0.5	0.34	0.16
x	12	14	16	18	20

16. Given the values of *w* and *x*, evaluate $\Sigma(w \cdot x)$.

w	0.3	0.19	0.17	0.16	0.24
x	25.4	24.9	25.67	24.18	24.87

17. Given the values of *z*, evaluate $\Sigma(z+3.5)$.

z	0	3	5	10	12	15

18. Given the values of *a* and *b*, evaluate $\Sigma(a-b)$.

a	8	4	19	22	15	9
b	2	4	6	8	10	12

19. Given the values of *z*, evaluate $\Sigma(z-1.5)$.

z	–6	–3	0	3	6	9

20. Given the values of *z*, evaluate $\Sigma(2a+b)$.

a	1	5	–2	7	–5	–4
b	0	3	–6	–5	–9	8

Chapter 2 Descriptive Statistics

Evaluating Formulas (Section 2.1)

Formula Notation

In algebra, you use variables such as *x*, *y*, and *a* to represent values in equations that change. In science, you use variables such as *F* and *C* to represent values in formulas that change. In algebraic equations, it is often the case that either *x* or *y* or any other letter could be used. In science formulas, using a letter other than *F* or *C* in the temperature conversion formula would not make sense because *F* stands for the Fahrenheit temperature and *C* stands for the Celsius temperature. Thus, the choice of letters is usually not optional.

In statistics, there are many formulas and there are many letters that have specific meanings. The meaning for the lowercase *n* is different from the meaning for the uppercase *N*, so you must be careful that you are using the correct one. Notice that the letters are typed in italics. Most mathematical software programs accommodate this.

Regardless if you are working with an equation in algebra class, a formula in science class, or a formula in statistics class, you must continue to work with the same letter or letters that are used in the original problem. You cannot change to your favorite letter(s), which for most students is *x*.

Manipulating Formulas

Formulas look like equations, in that they have an equal sign: =. Formulas have more than one variable. In statistics, the formula for the sample mean is $\bar{x} = \frac{\Sigma x}{n}$. Recall that $\bar{x}$ is read as "x-bar." When working with formulas, you can algebraically manipulate them to solve for other variables. In the formula for the sample mean, if you were to multiply each side by *n*, you would get $\bar{x}n = \Sigma x$. You have now solved for Σx. If you were to go a step further and divide each side by $\bar{x}$, you would get $n = \frac{\Sigma x}{\bar{x}}$. Now you have solved for *n*. Note: It is important to remember to write the bar over the *x*. There is a difference between *x* and $\bar{x}$.

Example 1: Solve for variable *a* in the formula $m = \frac{a+b}{2}$.

Solution: The variable *a* is in the numerator, so the first step is to isolate the numerator by eliminating the denominator. The next step is to isolate the term containing *a*.

$m = \frac{a+b}{2}$	Write original formula.
$m \cdot 2 = \frac{a+b}{2} \cdot 2$	Multiply to eliminate denominator.
$2m = a+b$	Simplify.
$2m - b = a + b - b$	Subtract to isolate the *a*-term.
$2m - b = a$	Simplify.
$a = 2m - b$	Rewrite.

Evaluating Formulas

When evaluating formulas, substitute values for the variables that you know to determine the value of the variable that you do not know. When substituting into an expression, it is wise to put parentheses around the values. If you know that $n = 40$ and $\bar{x} = 3.2$, you can use the formula $\bar{x} = \frac{\Sigma x}{n}$ to find Σx. After substituting, you get $3.2 = \frac{\Sigma x}{40}$. Multiplying each side by 40, you get $\Sigma x = 128$. You also could have used the rewritten version of the formula, $\bar{x}n = \Sigma x$, and arrived at the same answer: $(3.2)(40) = \Sigma x$.

There are two main approaches to evaluating formulas. Which one you use is personal preference.

- Substitute values for the variables that you know, and then determine the value of the unknown variable.
- Solve the formula for the unknown variable, and then substitute values for the variables that you know and evaluate.

Example: Substitute values for the variables that you know, and then determine the value of the unknown variable.

Solve for *m*: $y = mx + b$, $y = 7$, $x = -2$, $b = 4$.

Solution:

$y = mx + b$	Write original formula.
$7 = m(-2) + 4$	Substitute known values.
$7 - 4 = m(-2) + 4 - 4$	Subtract 4 from each side.
$3 = -2m$	Simplify.
$-\frac{3}{2} = m$	Divide each side by –2.
$m = -\frac{3}{2}$	Value of the unknown variable

Example: Solve the formula for the unknown variable, and then substitute values for the variables that you know and evaluate.

Solve for *m*: $y = mx + b$, $y = 7$, $x = -2$, $b = 4$.

Solution:

$y = mx + b$	Write original formula.
$y - b = mx + b - b$	Subtract to isolate the term containing *m*.
$y - b = mx$	Simplify.
$\frac{y-b}{x} = \frac{mx}{x}$	Divide to isolate *m*.
$\frac{y-b}{x} = m$	Divide out the common factor of *x*.
$m = \frac{y-b}{x}$	Solved for the unknown variable *m*.

Now substitute values for the variables that you know and evaluate.

$m = \frac{7-4}{-2}$ Substitute values for the known variables.

$m = \frac{3}{-2}$ Simplify the numerator.

$m = -\frac{3}{2}$ Value of the unknown variable

Example: Solve the formula for the unknown variable, and then substitute values for the variables that you know and evaluate.

Solve for n: $s = \frac{x}{n-1}$, $s = 11.2$, $x = 56$.

Solution:

$s = \frac{x}{n-1}$ Write original formula.

$s(n-1) = \frac{x}{n-1}(n-1)$ Multiply to eliminate the denominator.

$s(n-1) = x$ Simplify.

$\frac{s(n-1)}{s} = \frac{x}{s}$ Divide to isolate the factor containing n.

$n-1 = \frac{x}{s}$ Simplify.

$n = \frac{x}{s} + 1$ Add to isolate the unknown variable.

Now substitute values for the variables that you know and evaluate.

$n = \frac{56}{11.2} + 1$ Substitute values for the known variables.

$n = 6$ Evaluate.

Exercises

1. Solve for the variable p: $\mu = np$ (Recall that μ is the Greek letter mu.)

2. Solve for the variable w: $P = 2l + 2w$

3. Solve for the variable x: $z = \frac{x-\mu}{\sigma}$ (Recall that σ is the Greek letter lowercase sigma.)

4. Solve for the variable N: $\mu = \frac{\Sigma x}{N}$

5. Solve for the variable *f*: $p = 100\frac{f}{n}$

6. Solve for the variable *k*: $y = bk - M$

7. Solve for the variable *n*: $T = \frac{5r}{n-1}$

In Exercises 8–20, determine the value of the unknown variable.

8. Solve for *n*: $\hat{p} = \frac{x}{n}$, $\hat{p} = 0.125$, $x = 11$

9. Solve for *h*: $g = h(j+2)$, $j = 5.2$, $g = 36$

10. Solve for *A*: $P = A + B - 3AC$, $P = 25$, $B = 9$, $C = 2$

11. Solve for *y*: $z = 3w + x - 2y$, $x = -2$, $w = -5$, $z = 10$

12. Solve for *T*: $P = \frac{n}{T} \cdot 100$, $P = 62.5$, $n = 50$

13. Solve for *W*: $h = \frac{vW}{t}$, $h = 0.8$, $v = 2.5$, $t = 5$

14. Solve for μ: $z = \frac{x - \mu}{\sigma}$, $z = -0.5$, $x = 3.15$, $\sigma = 0.1$

15. Solve for *b*: $y = bk - M$, $y = -34$, $k = -12$, $M = -57.2$

16. Solve for *Q*: $t = pQR$, $t = 0.2106$, $p = -0.06$, $R = 2.7$

17. Solve for *r*: $T = \frac{5r}{n-1}$, $T = 0.1$, $n = 35$

18. Solve for g: $f = \frac{a+b}{g+h}$, $f = 1.25$, $a = 5$, $b = -25$, $h = 4$

19. Solve for *d*: $y = k(5-d)$, $y = -18$, $k = \frac{3}{2}$

20. Solve for *N*: $\mu = \frac{\Sigma x}{N}$, $\mu = 52.48$, $\Sigma x = 1312$

Chapter 2 Descriptive Statistics

Operations with Summations (Section 2.3)

Simple Operations with Summations

In subjects like statistics and calculus, summations are embedded within formulas. They fit into the order of operations as a group. The summation is calculated, and then the other operations are applied.

In the lesson on evaluating formulas, you looked at the formula for the sample mean, $\overline{x} = \frac{\Sigma x}{n}$, where n is the number of x values. To evaluate, you would take the sum of the x values, and then divide that sum by the value of n.

Example: Find the sample mean. Round to one decimal place.

x	23	40	65	52	38	41

Solution: The formula for the sample mean is $\overline{x} = \frac{\Sigma x}{n}$. Begin by finding the sum of the x-values.

$$\overline{x} = \frac{\Sigma x}{n}$$ Formula for sample mean

$$\overline{x} = \frac{23+40+65+52+38+41}{n}$$ Expand the summation.

$$\overline{x} = \frac{259}{n}$$ Add the numerator.

$$\overline{x} = \frac{259}{6}$$ $n = 6$

$$\overline{x} = 43.16666667$$ Divide.

$$\overline{x} \approx 43.2$$ Round to one decimal place.

Operations with More Than One Summation

Some formulas contain more than one summation. It does not matter which summation you evaluate first, but the summations should be evaluated first. Then the order of operations is followed.

Example: Evaluate $\frac{\Sigma(x \cdot w)}{\Sigma w}$. Round to one decimal place.

x	82	76	75	81	90
w	0.15	0.15	0.15	0.15	0.40

Solution: Begin by evaluating the summations.

$$\frac{\Sigma(x \cdot w)}{\Sigma w} = \frac{(82)(0.15)+(76)(0.15)+(75)(0.15)+(81)(0.15)+(90)(0.40)}{0.15+0.15+0.15+0.15+0.40}$$ Expand the summations.

$$= \frac{12.3+11.4+11.25+12.15+36}{0.15+0.15+0.15+0.15+0.40}$$ Multiply.

$$= \frac{83.1}{1} \quad \text{Add.}$$

$$= 83.1 \quad \text{Divide.}$$

Exercises

In Exercises 1–6, find the sample mean, $\bar{x} = \frac{\Sigma x}{n}$. Round to one decimal place.

1.

x	5	16	12	8	11	12	11	9

2.

x	19	–20	–21	18	–17	25

3.

x	–4.5	2.25	0	–1.75	1.25	2.5

4.

x	–72.4	–71.8	–68.4	–67.2	–71.5	–70.9	–72.1	–69.9

5.

x	154	156	155.2	154.3	155	155.5	154.2	154.1	154

6.

x	–0.2	–0.16	0.15	–0.19	0.09	–0.03	–0.01	0.04	0.05	0.17	–0.11

7. Evaluate $10\Sigma y$. Round to one decimal place.

y	8	11	15	10	9	9	8

8. Evaluate $-43\Sigma y$. Round to one decimal place.

y	–21	34	–41	–65	57	64	–72	55

9. Evaluate $\frac{2}{3}\Sigma y$. Round to one decimal place.

y	–178	–256	–224	–185	–193	–227	–203	–194

10. Evaluate $-0.078\Sigma y$. Round to one decimal place.

y	42	39	40	41	38	41	39	42	40	37	39

11. Evaluate $\frac{\Sigma z}{2}$. Round to one decimal place.

z	–1	0	1	–1	0	1	–1

12. Evaluate $\frac{\Sigma z}{16}$. Round to one decimal place.

z	–44	–67	–68	48	39	61	–72	–68	65	60

13. Evaluate $\frac{\Sigma z}{100}$. Round to one decimal place.

z	–2056	–2678	–2845	–2913	–2847	–2361	–2001	–2224

14. Evaluate $\frac{\Sigma z}{-6}$. Round to one decimal place.

z	14	16	19	24	28	37	39	42

15. Evaluate $\frac{\Sigma x}{\Sigma(x \cdot w)}$. Round to one decimal place.

x	79	65	80	68	73
w	0.25	0.25	0.20	0.20	0.10

16. Evaluate $\frac{\Sigma x}{\Sigma(x \cdot w)}$. Round to one decimal place.

x	345	361	355	339	370
w	0.20	0.20	0.20	0.20	0.20

17. Evaluate $\frac{\Sigma x}{\Sigma(x \cdot w)}$. Round to one decimal place.

x	1,000	100	10	1	–1
w	0.01	0.04	0.05	0.40	0.50

18. Evaluate $\frac{\Sigma(x \cdot w)}{(\Sigma x)(\Sigma w)}$. Round to two decimal places.

x	95	91	99	88	97	96
w	0.10	0.10	0.25	0.10	0.10	0.35

19. Evaluate $\frac{\Sigma(a \cdot b)}{\Sigma(a \cdot c)}$. Round to two decimal places.

a	0.2	0.1	0.4	0.2	0.1
b	4	2	3	1	4
c	10	9	7	9	8

20. Evaluate $\frac{\Sigma(a-b)}{\Sigma b}$. Round to two decimal places.

a	4.50	2.75	3.50	4.25	3.75
b	2	5	4	5	3

Chapter 2 Descriptive Statistics

Exponents (Section 2.4)

Base to a Power

When working with exponents, there is a base that is raised to a power. The power is the exponent.

It is important to determine what the base is, and to raise the entire base to the power. There is a difference between $(-5)^2$ and -5^2. In $(-5)^2$, the parentheses have priority over the exponent, so the base is -5 and the exponent is 2. The expression is evaluated as follows: $(-5)^2 = (-5)(-5) = 25$. In -5^2, the exponent has priority over the multiplication of the -1, so the base is 5 and the exponent is 2. The exponent applies only to the 5. The expression is evaluated as follows: $-5^2 = -(5)(5) = -25$.

Example: Evaluate $(-8)^2$ and -8^2.

Solution: $(-8)^2 = (-8)(-8) = 64$ — Parentheses have priority.

$-8^2 = -(8)(8) = -64$ — Exponent has priority.

Substituting in Formulas with Exponents

In the formula $f = n^2$, the base is *n* and the exponent is 2. Whatever the value of *n* is, it is squared. Because any real number squared is positive, *f* will always be positive. If $n = 7$, then $f = (7)^2 = 49$. If $n = -7$, then $f = (-7)^2 = 49$. Notice the use of parentheses when substituting values for *n*. Parentheses will help you correctly substitute negative values.

Example: Find $y = \frac{m}{x^2}$ when $m = 3$ and $x = -6$. Leave your answer in reduced fraction form.

Solution:

$$y = \frac{m}{x^2}$$ Write original formula.

$$y = \frac{3}{(-6)^2}$$ Substitute for *m* and *x*.

$$y = \frac{3}{36}$$ Parentheses have priority.

$$y = \frac{3 \cdot 1}{3 \cdot 12}$$ Factor out common factor.

$$y = \frac{\cancel{3} \cdot 1}{\cancel{3} \cdot 12}$$ Divide out common factor.

$$y = \frac{1}{12}$$ Simplify.

Example: Find $y = \Sigma(x-\mu)^2$ when $\mu = 6.5$ and the values of x are given in the following table. Find y to two decimal places.

x	4.1	5	7.2	6.7	6	10

Solution: The order of operations is parentheses, and then exponent. The summation is done at the end. When summations get more complicated, it can be helpful to work in the table itself, as follows.

x	4.1	5	7.2	6.7	6	10
$x-\mu$	4.1 – 6.5	5 – 6.5	7.2 – 6.5	6.7 – 6.5	6 – 6.5	10 – 6.5

Parentheses have first priority.

x	4.1	5	7.2	6.7	6	10
$x-\mu$	–2.4	–1.5	0.7	0.2	–0.5	3.5

Simplify.

x	4.1	5	7.2	6.7	6	10
$x-\mu$	–2.4	–1.5	0.7	0.2	–0.5	3.5
$(x-\mu)^2$	5.76	2.25	0.49	0.04	0.25	12.25

Exponents have second priority.

You now have all the values you need to find the summation.

$$y = \Sigma(x-\mu)^2$$ Write original formula.

$$= 5.76 + 2.25 + 0.49 + 0.04 + 0.25 + 12.25$$ Sum the values from the table.

$$= 21.04$$ Simplify.

Alternative Solution: Find the sum $y = (4.1-6.5)^2 + (5-6.5)^2 + (7.2-6.5)^2 + (6.7-6.5)^2 + (6-6.5)^2 + (10-6.5)^2$.

Exercises

In Exercises 1–8, evaluate the expression.

1. $(-9)^2$

2. -10^2

3. $-(4)^2$

4. $-(-1)^2$

5. -6^2

6. $(-13)^2$

7. $-\frac{1}{5^2}$

8. $\left(-\frac{1}{2}\right)^2$

9. Find $p = k^2$ when $k = 4$.

10. Find $q = d^2$ when $d = -8$.

11. Find $c = (w+2)^2$ when $w = 3$.

12. Find $k = (9 - w)^2$ when $w = 14$.

13. Find $y = \frac{p^2}{x}$ when $p = -4$ and $x = -48$. Leave your answer in reduced fraction form.

14. Find $y = \frac{p^2}{x}$ when $p = 10$ and $x = -15$. Leave your answer in reduced fraction form.

15. Find $y = \frac{x}{q^2}$ when $q = -6$ and $x = 10$. Leave your answer in reduced fraction form.

16. Find $q = r(p - 5)^2$ when $p = 8$ and $r = -4$.

17. Find $q = r(p - 5)^2$ when $p = 1$ and $r = \frac{1}{2}$.

18. Find $q = \frac{r}{(p + 2)^2}$ when $p = -10$ and $r = 36$. Leave your answer in reduced fraction form.

19. Find $q = r(p - 5)^2$ when $p = 3.4$ and $r = -0.42$. Round your answer to two decimal places.

20. Find $y = \Sigma(x - \mu)^2$ when $\mu = 12.1$ and the values of *x* are given in the following table. Round your answer to two decimal places.

x	13.4	10	15.6	12.6	11.2	9.8

Chapter 2 Descriptive Statistics

Operations with Square Roots (Section 2.4)

Simple Square Roots

The square root of a positive real number is always positive. In the last section, you learned that x^2 is positive for any real number. It is also true that $x = \sqrt{x^2}$ is positive for any positive real number. For example, $7 = \sqrt{7^2} = \sqrt{49}$.

It is true that you can square both 7 and –7 to get 49. However, the square root sign implies only a positive square root. This is known as the *principal square root.* Thus, the square root of 49 is positive 7, not negative 7: $\sqrt{49} = 7$ and $\sqrt{49} \neq -7$.

The square root of a negative number is not a real number, and will not be a topic in this book. For example, $\sqrt{-5}$ is not real.

In addition, note that if you square zero, you get $0^2 = 0$.

Example: Evaluate $-\sqrt{81}$.

Solution: There is a negative sign, but it is in front of the square root sign. Thus, it is implying multiplication by -1. You can think of a square root as a form of grouping. Thus, it will be performed before the multiplication.

$-\sqrt{81} = -9$ Simplify the square root.

Order of Operations with Square Roots

The square root sign, $\sqrt{\ }$, is a symbol for an exponent of $\frac{1}{2}$. You can write $\sqrt{25}$ as $25^{1/2}$. Because exponents have priority over multiplication and division, square roots also have priority over multiplication and division. Because grouping has priority over exponents, grouping also has priority over square roots.

PEMDAS is a popular acronym for the order of operations. P has the highest priority. There are six letters, but there are only four groups.

1. P: Parentheses—Some form of grouping, which may be implied grouping, such as a numerator.
2. E: Exponents—This includes square roots and other radicals, because they are symbolic exponents.
3. MD: Multiplication/Division—Equal priority; evaluated from left to right.
4. AS: Addition/Subtraction—Equal priority; evaluated from left to right.

Example: Evaluate $\sqrt{36}\sqrt{100}$.

Solution: The square roots have priority, and they can be evaluated.

$\sqrt{36}\sqrt{100} = 6\cdot 10$	Square roots have priority.
$= 60$	Multiply.

Example: Evaluate $\sqrt{3}\sqrt{3}$.

Solution: The square roots have priority, but they cannot be evaluated. Since both terms are square roots, they can be multiplied using the property $\sqrt{m}\sqrt{n} = \sqrt{mn}$.

$\sqrt{3}\sqrt{3} = \sqrt{3\cdot 3}$	$\sqrt{m}\sqrt{n} = \sqrt{mn}$
$= \sqrt{9}$	Multiply inside the square roots.
$= 3$	Evaluate the square root.

Example: Evaluate $\dfrac{\sqrt{125}}{\sqrt{5}}$.

Solution: The square roots have priority, but they cannot be evaluated. Because both terms are square roots, they can be divided using the property $\dfrac{\sqrt{m}}{\sqrt{n}} = \sqrt{\dfrac{m}{n}}$.

$\dfrac{\sqrt{125}}{\sqrt{5}} = \sqrt{\dfrac{125}{5}}$	$\dfrac{\sqrt{m}}{\sqrt{n}} = \sqrt{\dfrac{m}{n}}$
$= \sqrt{25}$	Divide inside the square roots.
$= 5$	Evaluate the square root.

Simplifying Square Roots

A square root is in simplest form if there are no factors that are perfect squares. For example, $\sqrt{12}$ is not in simplest form because 4 is a factor of 12, and 4 is a perfect square. To simplify, begin by factoring, $\sqrt{12} = \sqrt{4\cdot 3}$. Now apply the property $\sqrt{m}\sqrt{n} = \sqrt{mn}$ in reverse, giving $\sqrt{4\cdot 3} = \sqrt{4}\sqrt{3}$. Now you can evaluate the square roots, if possible $\sqrt{4}\sqrt{3} = 2\sqrt{3}$. Because no further simplification can be done, this is the simplest form. You are expected to simplify all square roots to the simplest form.

Example: Evaluate $\sqrt{10}\sqrt{6}$ and leave the answer in simplest form.

Solution:

$\sqrt{10}\sqrt{6} = \sqrt{10\cdot 6}$	$\sqrt{m}\sqrt{n} = \sqrt{mn}$
$= \sqrt{60}$	Multiply inside the square roots.
$= \sqrt{4\cdot 15}$	Factor out perfect square.
$= \sqrt{4}\sqrt{15}$	$\sqrt{m}\sqrt{n} = \sqrt{mn}$
$= 2\sqrt{15}$	Simplify the first square root.

Rationalizing the Denominator with Square Roots

Another requirement for square roots is that there are no square roots in the denominator. If a square root cannot be simplified, then it is an irrational number (a never-ending, never-repeating decimal number). An example of an irrational number is $\sqrt{3}$. The process of removing a square root from the denominator is called *rationalizing the denominator*. In this section, the properties $\sqrt{x}\sqrt{x} = \sqrt{x^2}$, $x = \sqrt{x^2}$, and $\frac{\sqrt{x}}{\sqrt{x}} = 1$ (when x is a positive real number) are used to rationalize the denominator.

Example: Rationalize the denominator $\frac{3}{\sqrt{6}}$ and leave the answer in simplest form.

Solution: The square root is in simplest form.

$\frac{3}{\sqrt{6}} = \frac{3}{\sqrt{6}} \cdot \frac{\sqrt{6}}{\sqrt{6}}$	Multiply by 1: $\frac{\sqrt{x}}{\sqrt{x}} = 1$.
$= \frac{3\sqrt{6}}{\sqrt{6}\sqrt{6}}$	Multiply fractions.
$= \frac{3\sqrt{6}}{\sqrt{36}}$	$\sqrt{x}\sqrt{x} = \sqrt{x^2}$
$= \frac{3\sqrt{6}}{6}$	$x = \sqrt{x^2}$
$= \frac{3\sqrt{6}}{3 \cdot 2}$	Factor out common factor.
$= \frac{\cancel{3}\sqrt{6}}{\cancel{3} \cdot 2}$	Divide out common factor.
$= \frac{\sqrt{6}}{2}$	Simplify.

Exercises

In Exercises 1–8, evaluate the square root.

1. $\sqrt{121}$ **2.** $\sqrt{0}$ **3.** $-\sqrt{144}$ **4.** $\sqrt{-25}$

5. $\sqrt{2}\sqrt{18}$ **6.** $\sqrt{20}\sqrt{5}$ **7.** $\frac{\sqrt{48}}{\sqrt{3}}$ **8.** $\sqrt{\frac{1}{4}}$

In Exercises 9–12, simplify the square root.

9. $\sqrt{63}$ **10.** $\sqrt{200}$

11. $\sqrt{80}$ **12.** $\sqrt{108}$

In Exercises 13–16, evaluate the square root and leave the answer in simplest form.

13. $\dfrac{\sqrt{180}}{\sqrt{2}}$

14. $\sqrt{15}\sqrt{10}$

15. $\sqrt{7}\sqrt{21}$

16. $\dfrac{\sqrt{88}}{\sqrt{11}}$

In Exercises 17–20, rationalize the denominator and leave the answer in simplest form.

17. $\dfrac{1}{\sqrt{3}}$

18. $\dfrac{2}{\sqrt{10}}$

19. $\dfrac{6}{\sqrt{15}}$

20. $\dfrac{\sqrt{3}}{\sqrt{6}}$

Putting It All Together

A formula used in statistics is $\sigma = \sqrt{\frac{\Sigma(x-\mu)^2}{N}}$. It contains many of the mathematical topics that you have been learning about—Greek letters, order of operations, fractions, summation, exponents, and square roots.

The Greek letters are lowercase sigma σ and mu μ. Notice that N is uppercase. According to PEMDAS, parentheses (or grouping) is performed first. The obvious grouping is $(x-\mu)$, which is included in the summation. The implied grouping is the numerator.

You could expand the summation, or you could utilize a table. For example, given $\mu = 24$, $N = 6$, and the given x-values, expand the table as follows.

x	23.8	24.1	25.6	21.3	23.9	25.3

→

x	23.8	24.1	25.6	21.3	23.9	25.3
$x-\mu$	–0.2	0.1	1.6	–2.7	–0.1	1.3

The formula is now $\sigma = \sqrt{\frac{(-0.2)^2+(0.1)^2+(1.6)^2+(-2.7)^2+(-0.1)^2+(1.3)^2}{6}}$. There is still implied grouping in the numerator, so that is the area of concentration. In the numerator, exponentiation has priority. You can work in the formula or in the table. The formula simplifies as $\sigma = \sqrt{\frac{0.04+0.01+2.56+7.29+0.01+1.69}{6}}$ and the table will expand as shown below.

x	23.8	24.1	25.6	21.3	23.9	25.3
$x-\mu$	–0.2	0.1	1.6	–2.7	–0.1	1.3
$(x-\mu)^2$	0.04	0.01	2.56	7.29	0.01	1.69

You can now find the value of the numerator, $\sigma = \sqrt{\frac{11.6}{6}}$. If the answer is to remain in square root form, you must rationalize the denominator, $\sigma = \sqrt{\frac{11.6}{6}\cdot\frac{6}{6}} = \sqrt{\frac{69.6}{36}} = \frac{\sqrt{69.6}}{6}$. If an approximate answer is required rounded to two decimal places, the implied grouping of the fraction has priority over the square root, $\sigma = \sqrt{\frac{11.6}{6}} = \sqrt{(11.6/6)} \approx 1.39$.

Exercises

In Exercises 1–8, simplify and (a) state the answer in square root form and (b) approximate the answer to two decimal places.

1. $\sigma = \sqrt{\frac{(1.5)^2+(-2.1)^2+(-0.3)^2+(1.4)^2+(1.9)^2}{5}}$

2. $\sigma = \sqrt{\dfrac{(-1.3)^2 + (1.2)^2 + (-0.7)^2 + (1.1)^2 + (-1.6)^2}{5}}$

3. $\sigma = \sqrt{\dfrac{(-2.5)^2 + (-1.8)^2 + (0.7)^2 + (-1.5)^2 + (0.8)^2 + (3.3)^2 + (-2.1)^2 + (0.2)^2}{8}}$

4. $s = \sqrt{\dfrac{(-0.7)^2 + (1.1)^2 + (-1.9)^2 + (-1.5)^2 + (0.9)^2 + (-1.7)^2}{6-1}}$

5. $s = \sqrt{\dfrac{(-1.4)^2 + (1.6)^2 + (0.6)^2 + (-1.8)^2 + (-0.3)^2 + (-1.1)^2}{6-1}}$

6. $s = \sqrt{\dfrac{(0.1)^2 + (1.5)^2 + (1.9)^2 + (-0.2)^2 + (-1.3)^2 + (-1.9)^2 + (1.2)^2 + (0.7)^2}{8-1}}$

7. $s = \sqrt{\dfrac{(-0.5)^2 + (0.2)^2 + (-1)^2 + (-0.1)^2 + (0)^2 + (-1.6)^2 + (0.2)^2 + (0.5)^2 + (-0.3)^2}{9-1}}$

8. $s = \sqrt{\dfrac{(0.3)^2 + (-0.2)^2 + (1.2)^2 + (-0.6)^2 + (-1.7)^2 + (-2.1)^2 + (0.4)^2 + (1.3)^2 + (-1.1)^2}{9-1}}$

9. Evaluate $\sigma = \sqrt{\dfrac{\Sigma(x-\mu)^2}{N}}$ given $\mu = 13.44$, $N = 5$, and the x-values in the table. Approximate the answer to two decimal places.

x	12.9	14.7	13.6	11.8	14.2

10. Evaluate $\sigma = \sqrt{\dfrac{\Sigma(x-\mu)^2}{N}}$ given $\mu = 8.18$, $N = 6$, and the x-values in the table. Approximate the answer to two decimal places.

x	8.9	9.5	7.7	6.4	8.4	8.2

11. Evaluate $\sigma = \sqrt{\dfrac{\Sigma(x-\mu)^2}{N}}$ given $\mu = 35.99$, $N = 9$, and the x-values in the table. Approximate the answer to two decimal places.

x	35.4	36.9	34.8	34.3	35	36.1	37.9	37.1	36.4

12. Evaluate $\sigma = \sqrt{\dfrac{\Sigma(x-\mu)^2}{N}}$ given $\mu = -10.36$, $N = 10$, and the x-values in the table. Approximate the answer to two decimal places.

x	−9.9	−9.5	−10.5	−11.2	−11.7	−10.8	−9.4	−9.8	−10.3	−10.5

13. Evaluate $s=\sqrt{\dfrac{\Sigma(x-\overline{x})^2}{n-1}}$ given $\overline{x}=31.34$, $n=5$, and the x-values in the table. Approximate the answer to two decimal places.

x	32.5	30.9	29.6	31.7	32

14. Evaluate $s=\sqrt{\dfrac{\Sigma(x-\overline{x})^2}{n-1}}$ given $\overline{x}=8.18$, $n=6$, and the x-values in the table. Approximate the answer to two decimal places.

x	8.9	9.5	7.7	6.4	8.4	8.2

15. Evaluate $s=\sqrt{\dfrac{\Sigma(x-\overline{x})^2}{n-1}}$ given $\overline{x}=-0.19$, $n=9$, and the x-values in the table. Approximate the answer to two decimal places.

x	–0.8	–0.3	0.1	–0.9	–0.8	–0.1	0.5	0.4	0.2

16. Evaluate $s=\sqrt{\dfrac{\Sigma(x-\overline{x})^2}{n-1}}$ given $\overline{x}=157.22$, $n=9$, and the x-values in the table. Approximate the answer to two decimal places.

x	157	160	154	155	161	150	162	157	159

17. Evaluate $d=\dfrac{(\Sigma x)^2}{\sqrt{n-1}}$ given $n=7$ and the x-values in the table. Approximate the answer to two decimal places.

x	4	3	8	7	1	0	5

18. Evaluate $d=\dfrac{(\Sigma x)^2}{\sqrt{n-1}}$ given $n=7$ and the x-values in the table. Approximate the answer to two decimal places.

x	15	21	19	25	31	27	36

19. Evaluate $d=\dfrac{(\Sigma x)^2}{\sqrt{n-1}}$ given $n=8$ and the x-values in the table. Approximate the answer to two decimal places.

x	1.2	1.4	1.8	1.4	1.5	1.3	1.2	1.1

20. Evaluate $d = \dfrac{(\Sigma x)^2}{\sqrt{n-1}}$ given $n = 10$ and the x-values in the table. Approximate the answer to two decimal places.

x	5	4.5	6	1.5	8	9.5	4.5	7	8.5	6

Chapter 2 Descriptive Statistics

Absolute Value (Section 2.4)

Most students recall that the **absolute value** of a real number is always positive, with the exception of zero (which is neither positive nor negative). For example, $|-3| = 3$ and $|0| = 0$. Why is absolute value always positive? Because absolute value stands for distance and distance is always positive. For any real number, the absolute value of the number represents its distance from zero on the real number line. The absolute value of -3 is equal to 3 because it is three units away from zero on the number line.

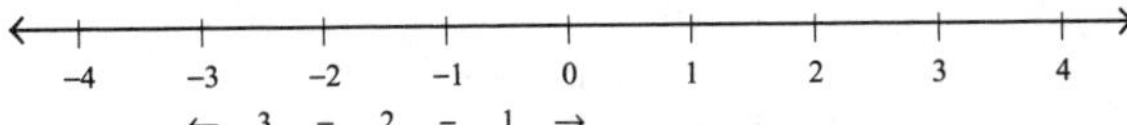

On the above number line, the number 3 is also three units away from zero. Thus, $|3| = 3$. It is correct to say that $|-3| = |3| = 3$, since both -3 and 3 are the same distance from zero on the number line. When this occurs, the pair of numbers are called **opposites.** Opposites are the same distance from zero on the number line, so they have the same absolute value. One is to the left of zero (negative) and the other is to the right of zero (positive). Opposites have opposite signs. To find the opposite of a number, multiply by -1.

Evaluating Absolute Values

The formal definition of absolute value involves opposites, as follows:

$$|x| = \begin{cases} x, & x \geq 0 \\ -x, & x < 0 \end{cases}$$

The absolute value of a positive number is itself. The absolute value of a negative number is its opposite.

Example: Evaluate $|0.26|$.

Solution: Since 0.26 is a positive number, the absolute value is itself: $|0.26| = 0.26$.

Example: Evaluate $|-0.94|$.

Solution: Since -0.94 is a negative number, the absolute value is its opposite: $|-0.94| = -(-0.94) = 0.94$.

Example: Evaluate $-|-1.4|$.

Solution: The absolute value is not affected by the leading negative sign. Since -1.4 is a negative number, the absolute value is its opposite: $-|-1.4| = -[-(-1.4)] = -[1.4] = -1.4$. The final answer is negative due to the leading negative sign in the original expression.

Example: Evaluate $\dfrac{|4.7-5.2|}{\sqrt{10-9}}$.

Solution: According to the order of operations, begin by simplifying the numerator and the denominator.

$$\frac{|4.7-5.2|}{\sqrt{10-9}}=\frac{|-0.5|}{\sqrt{1}}$$ Simplify inside the groupings.

$$=\frac{0.5}{\sqrt{1}}$$ Evaluate the absolute value.

$$=\frac{0.5}{1}$$ Evaluate the square root.

$$=0.5$$ Divide.

Exercises

In Exercises 1–6, evaluate the expression.

1. $|12|$
2. $|-4.5|$
3. $|-0.27|$
4. $-|0.051|$
5. $-|-0.94|$
6. $|0|$

In Exercises 7–17, evaluate the expression.

7. $0.25+|-0.25|$
8. $-0.8+|0.4|-|-0.9|-(-0.8)$
9. $|0.4|\cdot|-0.5|$
10. $-|-0.75|\cdot(0.6)\cdot|-0.45|$
11. $\frac{|-0.85|}{17}$
12. $-|-0.32|\div|0.04|$
13. $|.37|+0.84\div|-0.21|$
14. $-|0.741|-0.624\cdot|0.93|\div 0.2$
15. $-|0.21|+0.14\div|-0.7|\cdot 5-|0.62|$
16. $|-0.4|^2$
17. $\sqrt{\left|-\frac{1}{4}\right|}$
18. Evaluate $\Sigma|x|$.

x	0.1	–0.4	–0.3	0.7	0.9	–0.5

19. Evaluate $\Sigma|x|$.

x	-0.024	0.103	0.078	–0.45	–0.019	–0.026	0.027	0.011

20. Evaluate $\Sigma|x-0.05|$.

x	–0.03	0.28	0.34	–0.57	–0.26	0.79	0.18

Chapter 3 Probability

More Multiplication and Division (Section 3.1)

Repeated Multiplication Written as Exponents

Repeated multiplication of the same number is used in probability. An example is $8 \cdot 8 \cdot 8 \cdot 8 \cdot 8 \cdot 8 \cdot 8$. This repeated multiplication can be rewritten as 8^7, because 8 is a factor 7 times. The use of exponents simplifies your expression.

Example: Rewrite the repeated multiplications with exponents: $5 \cdot 3 \cdot 4 \cdot 4 \cdot 4 \cdot 7 \cdot 7 \cdot 7 \cdot 7 \cdot 7 \cdot 7$.

The 4 is repeated 3 times, and the 7 is repeated 6 times.

$$5 \cdot 3 \cdot 4^3 \cdot 7^6$$

Exact Answer vs. Approximate Answer

When division is involved, the final answer is often a fraction. You have learned how to rewrite a fraction in reduced form. In some areas of mathematics, it is better to present your answer in decimal form, rather than in fractional form. You use division to get from fractional form to decimal form. Sometimes you will get an exact answer and sometimes you will get an approximate answer.

For example, if $p = \frac{3}{8}$, then upon division, $p = 0.375$. If you do long division, the remainder is zero. If you use your calculator, there are only those three decimal digits. This is an exact answer. Thus, the use of an equal sign is appropriate.

However, if $p = \frac{3}{7}$, then long division does not leave a remainder of zero. The calculator gives you many digits that fill up the window. Depending on your calculator, you may see the answer as 0.4285714286. This is not an exact answer. If p is to be rounded to two decimal places, you may be tempted to write $p = 0.43$, but this is not correct because p is not exactly 0.43. You rounded to get 0.43. Therefore, 0.43 is an approximate answer, not an exact answer. The use of an equal sign is not appropriate. The use of the approximately equal to sign, $\approx$, is appropriate. So, write $p \approx 0.43$.

In conclusion, $p = \frac{3}{8}$ is an exact answer, and its decimal form, $p = 0.375$, is also an exact answer. By comparison, $p = \frac{3}{7}$ is an exact answer, and its decimal form, $p \approx 0.43$, is an approximate answer. A fraction is always an exact answer. Its decimal form is sometimes an exact answer.

Exercises

In Exercises 1–8, rewrite the repeated multiplications with exponents.

1. $1 \cdot 1 \cdot 1 \cdot 4 \cdot 5 \cdot 5 \cdot 5 \cdot 5 \cdot 5$
2. $3 \cdot 3 \cdot 4 \cdot 4 \cdot 4 \cdot 6 \cdot 9 \cdot 9 \cdot 9$
3. $5 \cdot 5 \cdot 5 \cdot 5 \cdot 5 \cdot 5 \cdot 11 \cdot 11 \cdot 11$
4. $3 \cdot 6 \cdot 6 \cdot 6 \cdot 10 \cdot 13 \cdot 13 \cdot 13 \cdot 13 \cdot 13 \cdot 17 \cdot 17$

5. $12 \cdot 12 \cdot 12 \cdot 12 \cdot 15 \cdot 15 \cdot 20 \cdot 20 \cdot 20 \cdot 21 \cdot 25 \cdot 27$

6. $4 \cdot 4 \cdot 4 \cdot 4 \cdot 4 \cdot 4 \cdot 4 \cdot 5 \cdot 9 \cdot 9 \cdot 11 \cdot 11 \cdot 11 \cdot 11 \cdot 11$

7. $56 \cdot 56 \cdot 56 \cdot 62 \cdot 62 \cdot 62 \cdot 62 \cdot 63 \cdot 63 \cdot 70 \cdot 70 \cdot 70 \cdot 70 \cdot 70$

8. $100 \cdot 100 \cdot 100 \cdot 101 \cdot 101 \cdot 101 \cdot 101 \cdot 102 \cdot 102 \cdot 102 \cdot 102 \cdot 103 \cdot 104 \cdot 104 \cdot 104 \cdot 104 \cdot 104 \cdot 104 \cdot 104$

In Exercises 9–20, rewrite the fractional answer as a decimal answer. Be sure to use the correct symbol, = or ≈. If an approximate answer is given, round to two decimal places.

9. $p = \frac{5}{8}$

10. $r = \frac{2}{13}$

11. $t = \frac{1}{6}$

12. $c = \frac{9}{5}$

13. $w = \frac{38}{3}$

14. $p = \frac{2}{100}$

15. $s = \frac{19}{1000}$

16. $h = \frac{154}{100,000}$

17. $v = \frac{18}{63}$

18. $q = \frac{2}{91}$

19. $b = \frac{254}{541}$

20. $r = \frac{65}{80}$

Chapter 3 Probability

Double Inequalities (Section 3.1)

In their basic form, double inequalities contain two inequality symbols with a variable between them and numbers outside of them. The inequality symbols are either less than or equal to symbols, $\leq$, or less than symbols, <. An example is $-2 \leq P \leq 1$. To graph this inequality on a number line, use square brackets at -2 and 1. The square brackets indicate that -2 and 1 are possible values for *P*. The square brackets open to the inside to indicate that every value in-between -2 and 1 is a possible value for *P*.

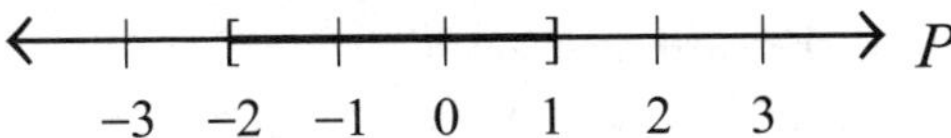

Example: Graph the double inequality $-3 < x < 3$ on the number line.

Solution: The double inequality has two less than symbols, so the number line will have two parentheses rather than two square brackets. The parentheses indicate that -3 and 3 are not possible values for *x*. The parentheses open to the inside to indicate that every value in-between -3 and 3 is a possible value for *x*.

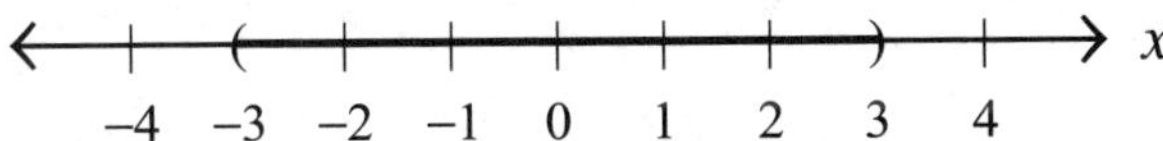

Example: Graph the double inequality $-0.5 < z \leq 1.5$ on the number line.

Solution: The double inequality will have a parenthesis on the left (<) and a square bracket on the right ($\leq$). The parenthesis indicates that -0.5 is not a possible value for *z*. The square bracket indicates that 1.5 is a possible value for *z*. The parenthesis and square bracket will open to the inside to indicate that every value in-between -0.5 and 1.5 is a possible value for *z*.

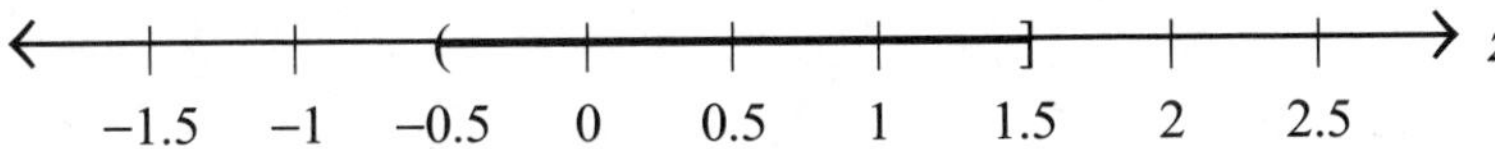

Rewrite a Double Inequality

When a double inequality contains a variable expression between the two inequality symbols, the goal is to rewrite it in its basic form. The basic form has only the variable between the two inequality symbols. The inequality symbols contain less than symbols, not greater than symbols.

During the process, you may have a double inequality such as $5 \geq x \geq -2$. This double inequality is not in basic form. It also does not align with the number line. The number line has the lesser numbers to the left and the greater numbers to the right. The statement, "Esteban is older than Jim" is the same as saying, "Jim is younger than Esteban." Thus, $5 \geq -2$ is the same as saying $-2 \leq 5$. You reverse the inequality. So, $5 \geq x \geq -2$ can be rewritten as $-2 \leq x \leq 5$.

Recall that when working with inequalities, multiplying or dividing by a negative number reverses the inequality.

Example: Rewrite the double inequality in its basic form: $4 \leq x+5 \leq 9$.

Solution:	$4 \leq x+5 \leq 9$	Write original inequality.
	$4-5 \leq x+5-5 \leq 9-5$	Subtract 5 from all three parts.
	$-1 \leq x \leq 4$	Simplify.

Example: Rewrite the double inequality in its basic form: $-3 \le -2x \le 7$.

Solution:

$-3 \le -2x \le 7$	Write original inequality.
$\frac{-3}{-2} \ge \frac{-2x}{-2} \ge \frac{7}{-2}$	Divide all three regions by –2 and reverse the inequalities.
$\frac{3}{2} \ge x \ge -\frac{7}{2}$	Simplify.
$-\frac{7}{2} \le x \le \frac{3}{2}$	Reverse the inequality.

Exercises

In Exercises 1–8, graph the double inequality on the number line. Label the number line appropriately.

1. $-3 \le y \le 2$

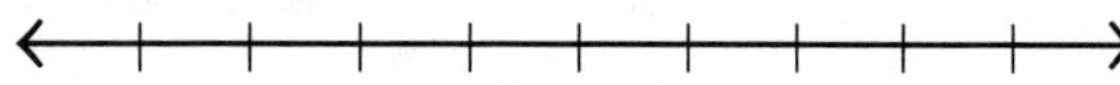

2. $0 < w < 5$

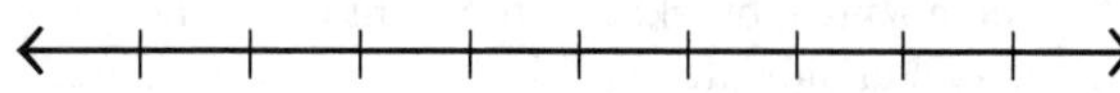

3. $-1 < t \le 1$

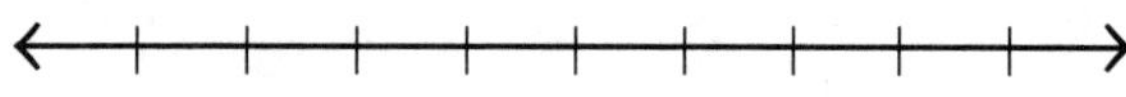

4. $1 \le z < 3$

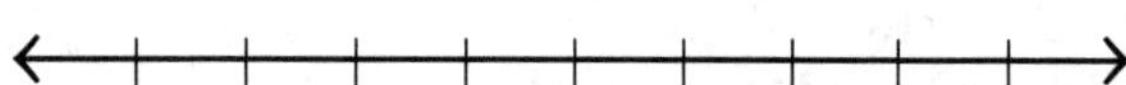

5. $0 \le P \le 1$

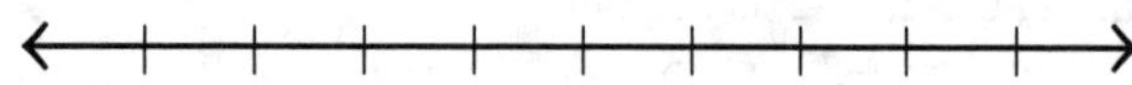

6. $-1.5 \le x < 2$

7. $-2.5 < z < -0.5$

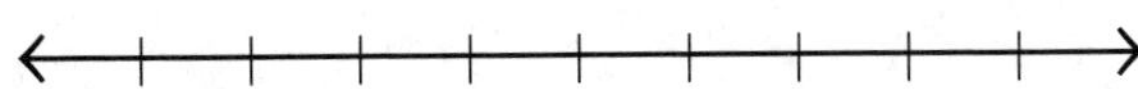

8. $0 < y \le 3.5$

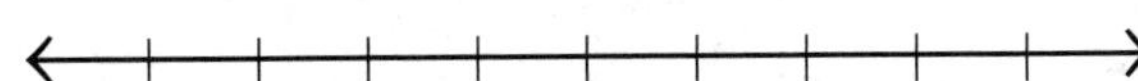

In Exercises 9–20, rewrite the double inequality in its basic form.

9. $2 \le x+8 \le 10$

10. $5 \le x-3 < 12$

11. $0 \le x-9 \le 20$

12. $-0.24 \le x-0.1 \le 0.24$

13. $-2.47 < x+1.75 < 2.47$

14. $-6 < 3x < 18$

15. $-10 \le -2z \le 16$

16. $1 < \frac{z}{2} < 5$

17. $-2 \le \frac{x-1}{3} \le 2$

18. $-8 < \frac{x+3}{2} < 8$

19. $-2.5 \le x-3 \le 2.5$

20. $-1.5 \le \frac{x-2}{3} \le 1.5$

Chapter 3 Probability

Translating Verbal Phrases (Section 3.2)

In this section, you will work through a variety of examples of verbal phrases and how they are translated mathematically. Then, you will have an opportunity to translate some on your own.

Example 1: Translate the following verbal phrase into a mathematical expression: The student's age, *a*, is between 18 and 22, inclusive.

Solution: The words "between 18 and 22" indicate an interval that would be translated as a double inequality. This phrase translates to $18 < a < 22$. However, the word "inclusive" indicates that the endpoints are to be included. So, the final translation is $18 \le a \le 22$.

Example 2: Translate the following verbal phrase into a mathematical expression: Thc price you will accept for your car, *p*, is at least \$1500.

Solution: The words "at least \$1500" indicate an inequality. On the number line, the least is to the left and the greatest is to the right. In this inequality, \$1500 is on the left, so \$1500 is a minimum value. There is no specific value on the right, so it approaches $+\infty$. So, the translation is $p \ge 1500$.

Example 3: Translate the following verbal phrase in to a mathematical expression: The number of grapefruits on the tree, *g*, is greater than 50.

Solution: The words "greater than 50" indicate an inequality. Unlike "at least," "greater than" does not include the value. This is the only difference between the two. So, the translation is $g > 50$.

NOTE: The words "at most" and "less than" work the same as "at least" and "greater than." The phrase "*w* is at most 10" translates to $w \le 10$ and the phrase "*w* is less than 10" translates to $w < 10$.

Example 4: Translate the following verbal phrase into a mathematical expression: The number of houses in the gated community, *h*, is exactly 75.

Solution: The words "is exactly" indicate equality. So, the translation is $h = 75$.

Example 5: Translate the following verbal phrase into a mathematical expression: The number of fish in the pond, *f*, is approximately 120.

Solution: The words "is approximately" indicate that this is an estimate. So, the translation is $f \approx 120$.

Example 6: Translate the following verbal phrase in to a mathematical expression: The number of times you rolled a two with a six-sided die, *n*, is not 5.

Solution: The words "is not" is shortcut for "is not equal to." So, the translation is $n \ne 5$.

Exercises

In Exercises 1–20, translate the verbal phrase into a mathematical expression.

1. The number of ducks in the pond, *d*, is between 12 and 18.
2. The speed of the wind, *w*, is at most 55 miles per hour.
3. The measurement of the particle, *m*, is exactly 5 millimeters.
4. The number of tickets sold, *s*, must be at least 100 for the show to go on.
5. The number of broken glasses in the shipment, *g*, is not zero.

6. The temperature today, t, is greater than 80°F.

7. The number of people at the concert in the park, p, is approximately 500.

8. The amount of gas in the tank, g, is less than 3 gallons.

9. The ages of the competitors, c, is between 16 and 78, inclusive.

10. The number of empty seats on the plane, s, is at most 3.

11. The amount of water in the cup, w, is approximately 0.32 cup.

12. The number of pictures in the box, p, is greater than 50.

13. The number of players on the field, P, is not 9.

14. The average weekly grocery bill, g, is between $\$150$ and $\$200$.

15. The cost to rent the hall for the banquet, c, is exactly $\$500$.

16. The number of sea turtles on the beach, t, is at least 15.

17. The sizes of the apartments in the building, s, are between 1600 and 2500 square feet, inclusive.

18. The number of hurricanes predicted this year, h, is less than 12.

19. The depth of the water, d, is approximately 76.28 feet.

20. The proportion of cans in the shipment that are damaged, p, is at most 0.18.

Chapter 3 Probability

Factorials (Section 3.4)

Definition of a Factorial

Factorials are often encountered when working with probability or summation. The exclamation mark, !, is the symbol for ***n* factorial.** $n!$ is defined as the product of the positive integers from n down to 1. For example, $4! = 4 \cdot 3 \cdot 2 \cdot 1 = 24$. Because n is a positive integer, the product is always positive. Factorials involve products, so the value of $n!$ increases quickly as n increases. For example, $10! = 10 \cdot 9 \cdot 8 \cdot 7 \cdot 6 \cdot 5 \cdot 4 \cdot 3 \cdot 2 \cdot 1 = 3{,}628{,}800$.

There is one exception to this definition. There is often a need for **zero factorial**, so zero factorial is defined as $0! = 1$.

Operations within a Factorial

Some formulas require an expression within the factorial. For example, $(n-r)!$ involves grouping, so the subtraction must be performed prior to the factorial.

Example: Evaluate $(n-r)!$ when $n = 9$ and $r = 2$.

Solution:

$(n-r)! = (9-2)!$	Substitute.
$= 7!$	Subtract.
$= 7 \cdot 6 \cdot 5 \cdot 4 \cdot 3 \cdot 2 \cdot 1$	Expand the factorial.
$= 5040$	Multiply.

Exercises

In Exercises 1–8, evaluate $n!$.

1. $3!$

2. $5!$

3. $8!$

4. $0!$

5. $9!$

6. $6!$

7. $1!$

8. $12!$

In Exercises 9–14, evaluate $(n-r)!$.

9. $n = 12$, $r = 11$

10. $n = 15$, $r = 2$

11. $n = 9$, $r = 4$

12. $n = 20$, $r = 20$

13. $n = 25$, $r = 12$

14. $n = 35$, $r = 24$

In Exercises 15–20, evaluate $(n-1)!$.

15. $n = 10$

16. $n = 13$

17. $n = 1$

18. $n = 6$

19. $n = 7$

20. $n = 14$

Chapter 3 Probability

Operations with Factorials (Section 3.4)

Multiplying Factorials

Once evaluated, a factorial is a value. When multiplying two factorials, you are multiplying two values. For example, $2!\cdot 4! = (2\cdot 1)\cdot(4\cdot 3\cdot 2\cdot 1) = 2\cdot 24 = 48$. Notice the order of operations. The factorials are a form of grouping, so they are evaluated first.

Example: Evaluate $3!\cdot 3!$.

Solution:

$3!\cdot 3! = (3\cdot 2\cdot 1)\cdot(3\cdot 2\cdot 1)$	Expand the factorials.
$= 6\cdot 6$	Simplify inside the parentheses.
$= 36$	Multiply.

Dividing Factorials

When dividing two factorials, you are dividing two values. If your instructor allows you to use a calculator, then you have discovered the ease of using the factorial key. If not, then you have discovered that factorials are a lot of work! When dividing factorials, you can use your skills in reducing fractions to reduce the work.

Example: Evaluate $\frac{5!}{3!}$.

Solution: You could evaluate this problem in the same way that the multiplication problem above was evaluated, $\frac{5!}{3!} = \frac{5\cdot 4\cdot 3\cdot 2\cdot 1}{3\cdot 2\cdot 1} = \frac{120}{6} = 20$. You can also divide out common factors to reduce the work.

$\frac{5!}{3!} = \frac{5\cdot 4\cdot 3\cdot 2\cdot 1}{3\cdot 2\cdot 1}$	Expand the factorials.
$= \frac{5\cdot 4\cdot \cancel{3}\cdot \cancel{2}\cdot \cancel{1}}{\cancel{3}\cdot \cancel{2}\cdot \cancel{1}}$	Divide out the common factors.
$= \frac{5\cdot 4}{1}$	Simplify.
$= \frac{20}{1}$	Multiply.
$= 20$	Divide.

Example: Evaluate $\frac{21!}{17!}$.

Solution: The factorial values are large in this problem, so dividing out common factors is the best approach.

$$\frac{21!}{17!} = \frac{21\cdot 20\cdot 19\cdot 18\cdot 17\cdot 16\cdot 15\cdot 14\cdot 13\cdot 12\cdot 11\cdot 10\cdot 9\cdot 8\cdot 7\cdot 6\cdot 5\cdot 4\cdot 3\cdot 2\cdot 1}{17\cdot 16\cdot 15\cdot 14\cdot 13\cdot 12\cdot 11\cdot 10\cdot 9\cdot 8\cdot 7\cdot 6\cdot 5\cdot 4\cdot 3\cdot 2\cdot 1}$$ Expand the factorials.

$$= \frac{21\cdot 20\cdot 19\cdot 18\cdot \cancel{17}\cdot \cancel{16}\cdot \cancel{15}\cdot \cancel{14}\cdot \cancel{13}\cdot \cancel{12}\cdot \cancel{11}\cdot \cancel{10}\cdot \cancel{9}\cdot \cancel{8}\cdot \cancel{7}\cdot \cancel{6}\cdot \cancel{5}\cdot \cancel{4}\cdot \cancel{3}\cdot \cancel{2}\cdot \cancel{1}}{\cancel{17}\cdot \cancel{16}\cdot \cancel{15}\cdot \cancel{14}\cdot \cancel{13}\cdot \cancel{12}\cdot \cancel{11}\cdot \cancel{10}\cdot \cancel{9}\cdot \cancel{8}\cdot \cancel{7}\cdot \cancel{6}\cdot \cancel{5}\cdot \cancel{4}\cdot \cancel{3}\cdot \cancel{2}\cdot \cancel{1}}$$ Divide out common factors.

$$= \frac{21\cdot 20\cdot 19\cdot 18}{1}$$ Simplify.

$$= \frac{143,640}{1}$$ Multiply.

$$= 143,640$$ Divide.

Multiplying and Dividing Factorials

When multiplying and dividing factorials, be sure to follow the order of operations.

Example: Evaluate $\frac{9!4!}{5!}$.

Solution: If using a calculator with a factorial key, evaluate the numerator, then divide by the denominator. If entering as one command, enter as $(9!4!)/5!$. Without a calculator, divide out common factors as follows.

$$\frac{9!4!}{5!} = \frac{(9\cdot 8\cdot 7\cdot 6\cdot 5\cdot 4\cdot 3\cdot 2\cdot 1)\cdot(4\cdot 3\cdot 2\cdot 1)}{5\cdot 4\cdot 3\cdot 2\cdot 1}$$ Expand the factorials.

$$= \frac{9\cdot 8\cdot 7\cdot 6\cdot 5\cdot 4\cdot 3\cdot 2\cdot 1\cdot 4\cdot 3\cdot 2\cdot 1}{5\cdot 4\cdot 3\cdot 2\cdot 1}$$ Multiply.

$$= \frac{9\cdot 8\cdot 7\cdot 6\cdot \cancel{5}\cdot \cancel{4}\cdot \cancel{3}\cdot \cancel{2}\cdot \cancel{1}\cdot 4\cdot 3\cdot 2\cdot 1}{\cancel{5}\cdot \cancel{4}\cdot \cancel{3}\cdot \cancel{2}\cdot \cancel{1}}$$ Divide out common factors.

$$= \frac{9\cdot 8\cdot 7\cdot 6\cdot 4\cdot 3\cdot 2\cdot 1}{1}$$ Simplify.

$$= \frac{72,576}{1}$$ Multiply.

$$= 72,576$$ Divide.

Example: Evaluate $\dfrac{16!}{10!6!}$.

Solution: If using a calculator with a factorial key, evaluate the denominator completely, and then divide the numerator by the resulting denominator value. If entering as one command, enter as 16!/(10!6!). Without a calculator, divide out common factors as follows.

$$\frac{16!}{10!6!} = \frac{16\cdot15\cdot14\cdot13\cdot12\cdot11\cdot10\cdot9\cdot8\cdot7\cdot6\cdot5\cdot4\cdot3\cdot2\cdot1}{(10\cdot9\cdot8\cdot7\cdot6\cdot5\cdot4\cdot3\cdot2\cdot1)\cdot(6\cdot5\cdot4\cdot3\cdot2\cdot1)}$$ Expand the factorials.

$$= \frac{16\cdot15\cdot14\cdot13\cdot12\cdot11\cdot10\cdot9\cdot8\cdot7\cdot6\cdot5\cdot4\cdot3\cdot2\cdot1}{10\cdot9\cdot8\cdot7\cdot6\cdot5\cdot4\cdot3\cdot2\cdot1\cdot6\cdot5\cdot4\cdot3\cdot2\cdot1}$$ Multiply.

$$= \frac{16\cdot15\cdot14\cdot13\cdot12\cdot11\cdot\cancel{10}\cdot\cancel{9}\cdot\cancel{8}\cdot\cancel{7}\cdot\cancel{6}\cdot\cancel{5}\cdot\cancel{4}\cdot\cancel{3}\cdot\cancel{2}\cdot\cancel{1}}{\cancel{10}\cdot\cancel{9}\cdot\cancel{8}\cdot\cancel{7}\cdot\cancel{6}\cdot\cancel{5}\cdot\cancel{4}\cdot\cancel{3}\cdot\cancel{2}\cdot\cancel{1}\cdot6\cdot5\cdot4\cdot3\cdot2\cdot1}$$ Divide out common factors.

$$= \frac{16\cdot15\cdot14\cdot13\cdot12\cdot11}{6\cdot5\cdot4\cdot3\cdot2\cdot1}$$ Simplify.

$$= \frac{16\cdot15\cdot14\cdot13\cdot\cancel{12}\cdot11}{\cancel{6}\cdot5\cdot4\cdot3\cdot\cancel{2}\cdot1}$$ Divide out common factor 12.

$$= \frac{16\cdot15\cdot14\cdot13\cdot11}{5\cdot4\cdot3\cdot1}$$ Simplify.

$$= \frac{16\cdot\cancel{15}\cdot14\cdot13\cdot11}{\cancel{5}\cdot4\cdot\cancel{3}\cdot1}$$ Divide out common factor 15.

$$= \frac{16\cdot14\cdot13\cdot11}{4\cdot1}$$ Simplify.

$$= \frac{4\cdot4\cdot14\cdot13\cdot11}{4\cdot1}$$ Factor 16.

$$= \frac{\cancel{4}\cdot4\cdot14\cdot13\cdot11}{\cancel{4}\cdot1}$$ Divide out common factor 4.

$$= \frac{4\cdot14\cdot13\cdot11}{1}$$ Simplify.

$$= \frac{8008}{1}$$ Multiply.

$$= 8008$$ Divide.

Exercises

In Exercises 1–20, evaluate the expression. Leave your answer in fractional form, if applicable.

1. $1! \cdot 3!$

2. $2! \cdot 5!$

3. $4! \cdot 4!$

4. $6! \cdot 2!$

5. $\frac{4!}{3!}$

6. $\frac{5!}{8!}$

7. $\frac{7!}{6!}$

8. $\frac{23!}{20!}$

9. $\frac{13!}{12!}$

10. $\frac{18!}{14!}$

11. $\frac{14!}{5!}$

12. $\frac{5!}{3!6!}$

13. $\frac{5!3!}{2!}$

14. $\frac{4!2!}{2!}$

15. $\frac{19!}{13!6!}$

16. $\frac{6!}{5!1!}$

17. $\frac{25!}{22!3!}$

18. $\frac{8!}{4!4!}$

19. $\frac{14!}{9!5!}$

20. $\frac{2!5!}{3!6!}$

Chapter 3 Probability

Scientific Notation (Section 3.4)

Calculators are available on your computer and on most cell phones. Calculators have become familiar devices used every day. When you do calculations with large or small numbers, you probably use a calculator. However, sometimes the answer on the screen looks a little bit different depending on the calculator.

Find your favorite calculator and calculate $20!$. Your calculator screen may look something like this (with more or less digits): 2.432902008E18 or 2.432902008 18 or 2.432902008*10^18 or a different variation. Obviously $20!$ is larger than 20, so the 2.432902008 part alone cannot be the answer (since it is less than 3).

The calculator gave you the answer in scientific notation. When the answer has too many digits to be displayed in the calculator window, the answer is displayed in scientific notation. It is your responsibility to recognize your calculator's version of a number in scientific notation, and then rewrite it in the correct mathematical form of scientific notation. Many calculators do not display the number in the correct mathematical form of scientific notation.

The correct mathematical form of scientific notation for $20!$ is 2.432902008×10^{18}.

Multiplying by 10^{18} moves the decimal point 18 places to the right, resulting in a very large number: $2.432902008\times10^{18}=2,432,902,008,000,000,000$.

Look at the window of your calculator and notice how the 2.432902008 appears and how the 10^{18} appears. Now calculate $\frac{1}{20!}$. Your calculator window may display 4.11031762 E–19. That number in scientific notation is 4.11031762×10^{-19}. Multiplying by 10^{-19} moves the decimal point 19 places to the left, resulting in a very small number:
$4.11031762\times10^{-19}=0.000000000000000000411031762$.

Mathematical Form of a Scientific Notation Number

You have learned the correct form of scientific notation for very large and very small numbers, 2.432902008×10^{18} and 4.11031762×10^{-19}. They both consist of two factors, connected with the symbol $\times$ for multiplication.

The first factor contains decimals, but it does not have to. However, there is one and only one digit to the left of the decimal point, and it must be a digit from 1 to 9. Zero is not allowed. Another way to look at it is: $1\le$ first factor <10.

The second factor is 10 raised to a power. The exponent must be a whole number (no fractions or decimals). If the exponent is positive, the answer is large. If the exponent is negative, the answer is small (close to zero).

Example: Given the calculator answer, 5.06781524E14, write the number in the correct mathematical form of scientific notation, and indicate if the answer is large or small.

Solution: The correct mathematical form of the number in scientific notation is 5.06781524×10^{14}. Since the exponent is positive, the answer is large.

Example: Given the calculator answer, 7.34570001E–9, write the number in the correct mathematical form of scientific notation. Round the first factor to three significant digits.

Solution: The correct mathematical form of the number in scientific notation is 7.34570001×10^{-9}. Rounded to three significant digits, the answer is 7.35×10^{-9}.

Exercises

In Exercises 1–10, write the calculator answer in the correct mathematical form of scientific notation. Indicate if the answer is large or small.

1. 6.218547E8

2. 7.3900045E–12

3. 5E–14

4. 1.2500649E22

5. 3.995475E18

6. –6.425137E27

7. –9.5000024E–17

8. 2.0067818E21

9. –3E24

10. 8.2384579E–31

In Exercises 11-20, write the calculator answer in the correct mathematical form of scientific notation. Then round to three significant digits.

11. 3.82714E–9

12. 5.0071824E–18

13. 9.3627121E17

14. 6.00031892E13

15. 4.1182647E–25

16. 7.1553454E–10

17. 8.88964132E25

18. 1.91842374E37

19. 2.0032598E–31

20. 8.44494287E28

Chapter 4 Discrete Probability Distributions

More Translating Verbal Phrases (Section 4.2)

In this section, you will work through a variety of examples of verbal phrases and how they are interpreted mathematically. Then you will have an opportunity to interpret some on your own.

Example 1: Translate the following verbal phrase and interpret, letting n represent the value(s): "fewer than two people."

Solution: There cannot be a negative number of people. There could be zero people. There could be one person. There cannot be two people, because two is not fewer than two. There cannot be part of a person. There can only be a whole number of people. Therefore, $n = 0, 1$.

Example 2: Translate the following verbal phrase and interpret, letting n represent the value(s): "at least three men are in the room" (there are seven people in the room).

Solution: The minimum number of men is three. All the people could be men, which would be seven men. Any number in between would work—four, five, or six. There must be a whole number of men. Therefore, $n = 3, 4, 5, 6, 7$.

Example 3: Translate the following verbal phrase and interpret, letting n represent the value(s): "the tables that are occupied is between 2 and 6."

Solution: "Between" does not include the endpoints, so 3, 4, or 5 tables are occupied. A table is occupied, or it is not, so part of a table is not an answer. Therefore, $n = 3, 4, 5$.

Example 4: Translate the following verbal phrase and interpret, letting n represent the value(s): "two people or less."

Solution: There cannot be a negative number of people. There could be zero people. There could be one person. There could be two people. There must be a whole number of people. Therefore, $n = 0, 1, 2$.

Example 5: Translate the following verbal phrase and interpret, letting n represent the value(s): "two people or more."

Solution: There could be two people. There could be three people. There is no limit to the number of people. This is not an inequality, because you cannot represent people with a fractional number. There can only be whole number values. Therefore, $n = 2, 3, 4, 5, \ldots$.

Example 6: Translate the following verbal phrase and interpret, letting n represent the value(s): "at most two athletes will qualify."

Solution: The maximum number of athletes is two. There cannot be a negative number of athletes. There could be one athlete. There could be zero athletes. There must be a whole number of athletes. Therefore, $n = 0, 1, 2$.

Exercises

In Exercises 1–20, translate the verbal phrase and interpret, letting n represent the value(s).

1. "Three books or less"
2. "At most six people in the elevator"
3. "The ages of the children are between 4 and 9."
4. "Ten tickets or more"
5. "At least three women are in the room." (There are six people in the room.)
6. "Fewer than three cats"
7. "More than five dogs in the park." (There are eight dogs in the park.)

8. "The number of chairs per table is between 4 and 8, inclusive."

9. "Seven cars or more"

10. "Fewer than ten children"

11. "At most five ingredients in each recipe"

12. "Six lemons or less"

13. "The ages of the dogs are between 3 and 10 years, inclusive."

14. "At least four choices" (There are seven choices available.)

15. "More than twenty chairs in the room" (There are twenty-five chairs in the room.)

16. "The number of children per group is between 2 and 5."

17. "At most one person per room"

18. "Fewer than eight musicians"

19. "Fifty houses or more"

20. "Two occupants or less"

More Division Involving Square Roots

Using a Calculator

You learned how to perform division involving square roots without a calculator. In this section, you will concentrate on performing division involving square roots with a calculator to give a decimal answer to a given number of decimal places. A standard rule is to not round until the final answer. So, any intermediate results must be maintained to as many decimal places as displayed on your calculator window. The goal is to eliminate intermediate results.

Example 1: Evaluate $\sqrt{\frac{5}{2}}$. Round your answer to two decimal places.

Solution: The expression inside the square root, $\frac{5}{2}$, must be evaluated first. Think, $\sqrt{\left(\frac{5}{2}\right)}$. How you proceed from here depends on your calculator.

- Some calculators want the value and then the square root key.
    - Type $5 \div 2 =$. It is important to press the = key. You should see 2.5 in the calculator window. Now press the square root key.

      $$\sqrt{\frac{5}{2}} \approx 1.58113883$$

- Other calculators want the square root key and then the value. In this case, it is important to group what is inside the square root symbol by using parentheses.
    - Press the square root key. Type $(5 \div 2)$. Press the = or Enter key.

      $$\sqrt{\frac{5}{2}} \approx 1.58113883.$$

So, $\sqrt{\frac{5}{2}} \approx 1.58$ rounded to two decimal places.

Example 2: Evaluate $\frac{\sqrt{5}}{2}$. Round your answer to two decimal places.

Solution: Evaluate the numerator, $\sqrt{5}$, and then divide that value by 2. Be careful to complete the calculation of $\sqrt{5} \approx 2.236067977$ before dividing by 2.

$$\frac{\sqrt{5}}{2} \approx 1.118033989$$

So, $\frac{\sqrt{5}}{2} \approx 1.12$ rounded to two decimal places.

Example 3: Evaluate $\frac{3}{\sqrt{2}}$. Round your answer to two decimal places.

Solution: Because you want a decimal answer, you do not have to rationalize the denominator. How you proceed from here depends on your calculator.

- Some calculators want the value and then the square root key.
  - Type $3 \div 2\sqrt{\ } =$. It is important to remember to press the = key at the end.

    $\frac{3}{\sqrt{2}} \approx 2.121320344$

- Other calculators want the square root key and then the value. In this case, it is important to group what is inside the square root symbol by using parentheses.
  - Type $3 \div \sqrt{\ }(2)$. Press = or Enter.

    $\frac{3}{\sqrt{2}} \approx 2.121320344$

So, $\frac{3}{\sqrt{2}} \approx 2.12$ rounded to two decimal places.

Example 4: Evaluate $\frac{\sqrt{3}}{\sqrt{7}}$. Round your answer to two decimal places.

Solution: An efficient way to approach this expression is to use the property $\frac{\sqrt{m}}{\sqrt{n}} = \sqrt{\frac{m}{n}}$ to rewrite. Then, it can be evaluated using the method in Example 1. So, $\frac{\sqrt{3}}{\sqrt{7}} = \sqrt{\frac{3}{7}}$. Use the approach in Example 1.

$$\frac{\sqrt{3}}{\sqrt{7}} = \sqrt{\frac{3}{7}} = 0.6546536707$$

So, $\frac{\sqrt{3}}{\sqrt{7}} = 0.65$ rounded to two decimal places.

Exercises

In Exercises 1–20, evaluate the expression. Round your answer to two decimal places.

1. $\sqrt{\frac{3}{8}}$

2. $\sqrt{\frac{9}{5}}$

3. $\sqrt{\frac{1}{11}}$

4. $\sqrt{\frac{10}{3}}$

5. $\sqrt{\frac{21}{10}}$

6. $\sqrt{\frac{43}{13}}$

7. $\frac{\sqrt{6}}{3}$

8. $\frac{\sqrt{10}}{7}$

9. $\frac{\sqrt{5}}{4}$

10. $\frac{\sqrt{41}}{17}$

11. $\frac{\sqrt{26}}{3}$

12. $\frac{\sqrt{70}}{7}$

13. $\frac{4}{\sqrt{10}}$

14. $\frac{11}{\sqrt{13}}$

15. $\frac{9}{\sqrt{15}}$

16. $\frac{3}{\sqrt{29}}$

17. $\frac{\sqrt{15}}{\sqrt{2}}$

18. $\frac{\sqrt{5}}{\sqrt{17}}$

19. $\frac{\sqrt{28}}{\sqrt{3}}$

20. $\frac{\sqrt{11}}{\sqrt{22}}$

Chapter 6 Confidence Intervals

Intervals on the Number Line (Section 6.1)

There are many instances where a range or interval of values is given, rather than an exact value. For example, you bring your car to the shop for repair and you are told that it will cost between $600 and $800 to repair your car. The mechanic estimates that it will cost $700 to fix your car, and is allowing $100 either way. Mathematically, this could be written as $700 ± $100. The ± symbol implies two operations: addition and subtraction. It is a shortcut to writing $700 + $100 and $700 − $100. These two operations generate the estimates $800 and $600. In this scenario, the endpoints, $600 and $800, would be included in the interval. In other scenarios, they are not. In the graph, the endpoints are square brackets, just like the graph of a double inequality. The center of the interval is a point on the graph. The center is equidistant from both endpoints. On this graph, the center is $100 from each endpoint.

Example: Graph the interval on the number line. Include the endpoints and label the graph accordingly.

12 ± 3

Solution: The center is at 12. The endpoints are $12 - 3 = 9$ and $12 + 3 = 15$. The endpoints are included, so the endpoints are square brackets.

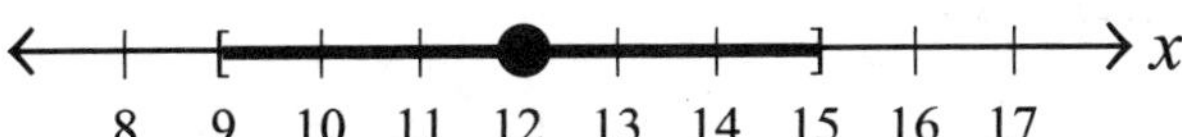

Example: Graph the interval on the number line. Do not include the endpoints and label the graph accordingly.

0 ± 50

Solution: The center is at 0. The endpoints are $0 - 50 = -50$ and $0 + 50 = 50$. The endpoints are not included, so the endpoints are parentheses.

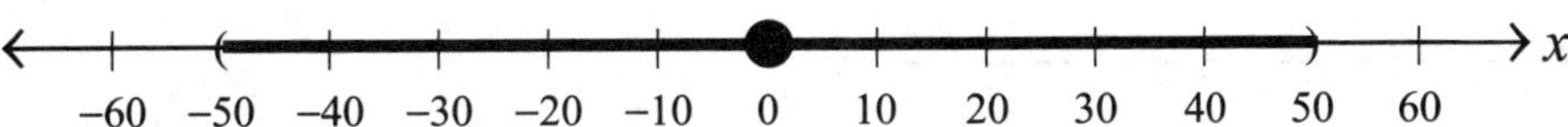

Exercises

In Exercises 1–10, graph the interval on the number line. Include the endpoints and label the graph accordingly.

1. 4 ± 1

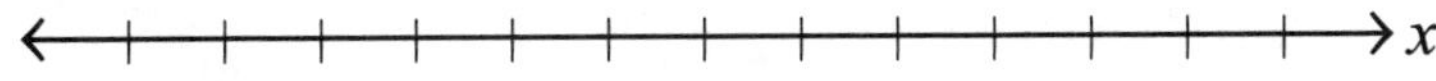

2. 0 ± 3

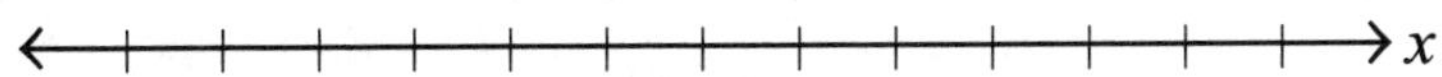

3. -2 ± 5

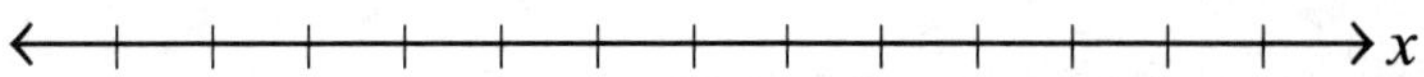

4. 10 ± 20

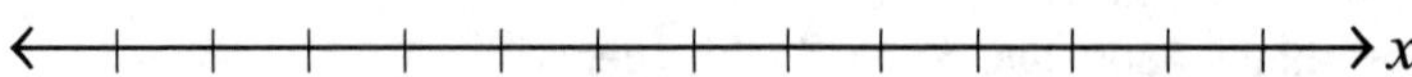

5. 300 ± 50

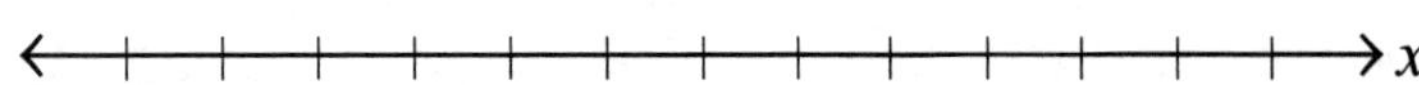

6. 0 ± 0.5

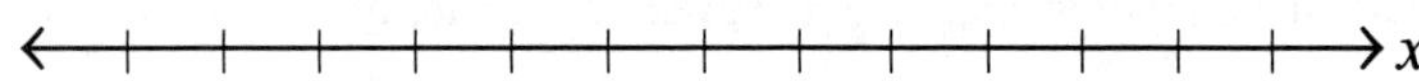

7. 3 ± 1.5

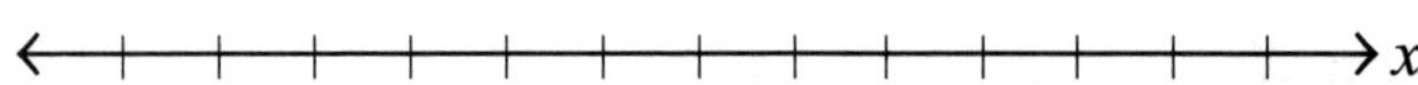

8. -1 ± 2.5

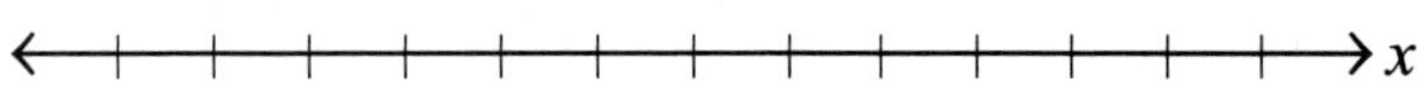

9. 1.5 ± 1

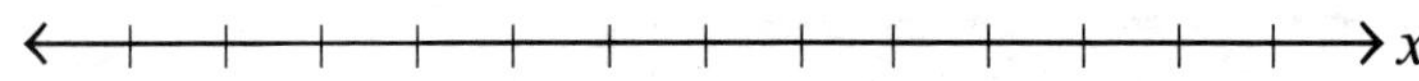

10. -1.5 ± 0.5

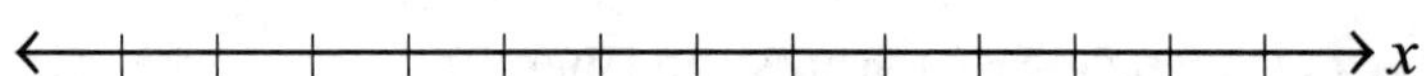

In Exercises 11–20, graph the interval on the number line. Do not include the endpoints and label the graph accordingly.

11. -5 ± 2

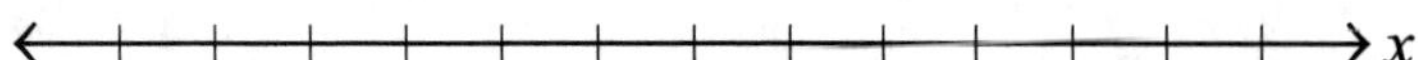

12. 12 ± 6

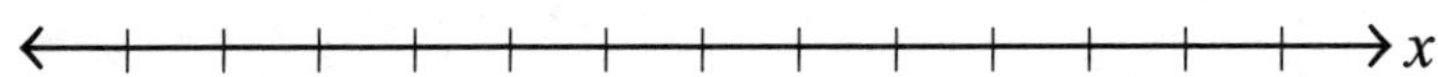

13. 0 ± 40

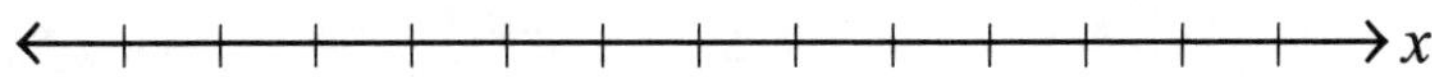

14. -20 ± 10

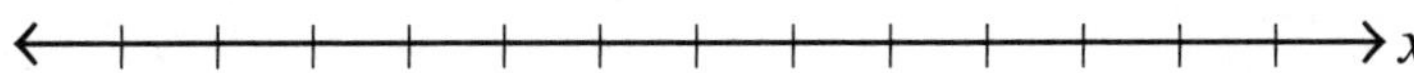

15. 4000 ± 1000

16. 6 ± 2.5

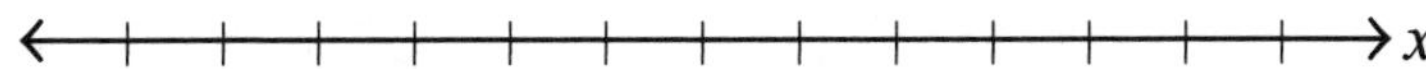

17. 0 ± 4.5

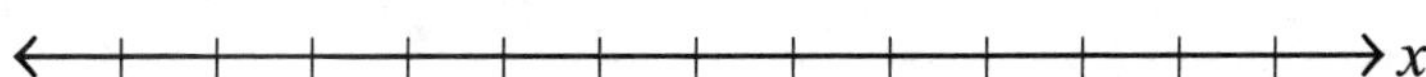

18. -3 ± 1.5

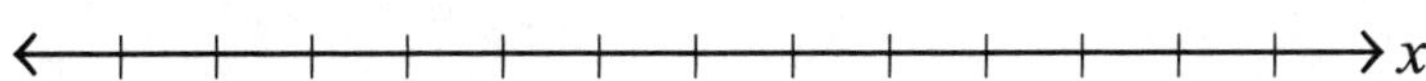

19. 2 ± 2.5

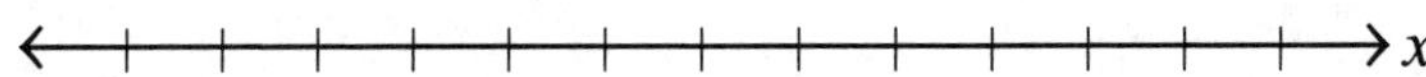

20. 1.5 ± 1.5

Chapter 9 Correlation and Regression

Slope-Intercept Form of the Equation of a Line (Section 9.2)

One form of the equation of a line that is useful for graphing is the slope-intercept form, $y = mx + b$, where m is the slope of the line and the point $(0, b)$ is the y-intercept. The variables x and y represent the infinitely many points (x, y) on the line, and each of these points is a solution of the equation.

In algebra, you concentrated on the slope and the y-intercept to graph the line. This is appropriate when the slope has values like $\frac{2}{5}$ or 3, where it is easy to visualize rise over run. In statistical research, the slope may have a value such as 12.51684. It is difficult to visualize rise over run to graph the slope. It may also be the case that the data values are concentrated on the x-axis around $10{,}000$. The y-intercept will require you to extend the graph far to the left (this will make more sense later when you are working with statistical research).

These complications often require a change in how you graph the line. The simplest approach is to plot two points. When you did this in algebra, you chose 0 and 1 values for x, but like the y-intercept, these values are probably not where your data is concentrated. Therefore, you pick two x-values within the concentrated data, solve for the corresponding y-values, plot the two points, and then draw the line between the points.

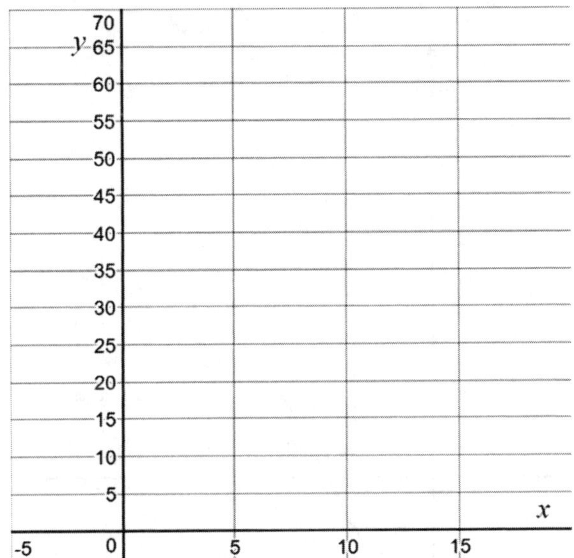

Example: Graph the line $y = 2.4x + 14.5$ on the graph.

Solution: The x-axis is from -5 to 20, so choose x-values such as 5 and 10 (these are values between -5 and 20).

When $x = 5$, then $y = 2.4(5) + 14.5 = 26.5$. Therefore, the point $(5, 26.5)$ is on the line.

When $x = 10$, then $y = 2.4(10) + 14.5 = 38.5$. Therefore, the point $(10, 38.5)$ is on the line.

Plot the points and draw the line through the two points.

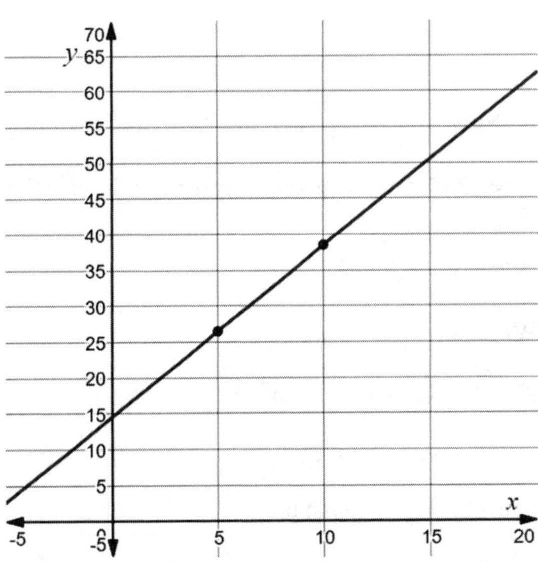

Example: Graph the line $y = 0.657x - 5.34$ on the graph.

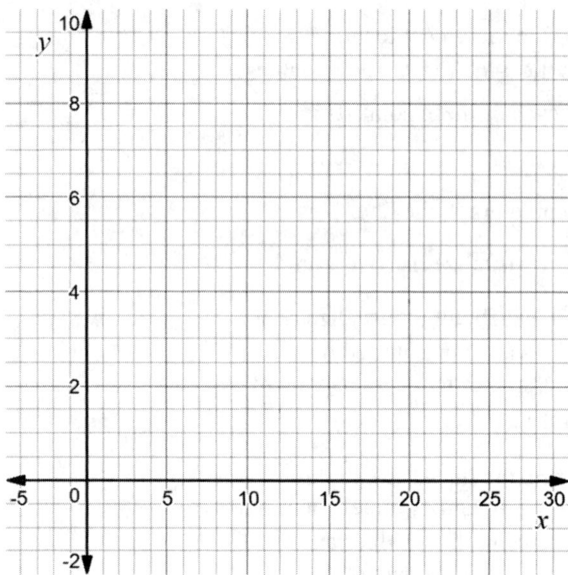

Solution: The x-axis is from -5 to 30, so choose x-values such as 10 and 20.

When $x = 10$, then $y = 0.657(10) - 5.34 = 1.23$. Therefore, the point $(10, 1.23)$ is on the line.

When $x = 20$, then $y = 0.657(20) - 5.34 = 7.8$. Therefore, the point $(20, 7.8)$ is on the line.

Plot the points and draw the line through the two points.

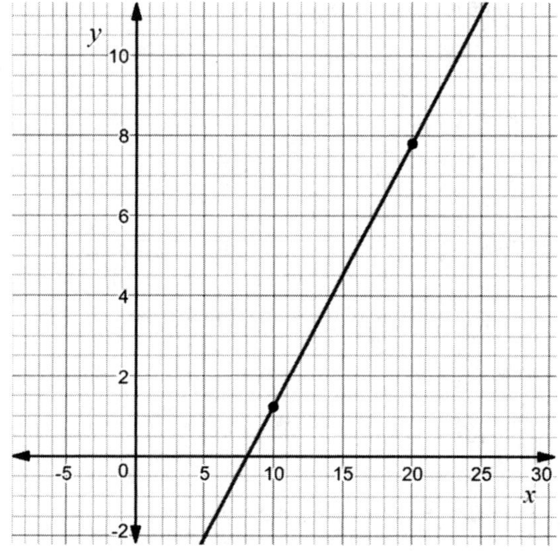

Exercises

In Exercises 1–4, use the given graph and the equation $y = 3.65x - 1.82$.

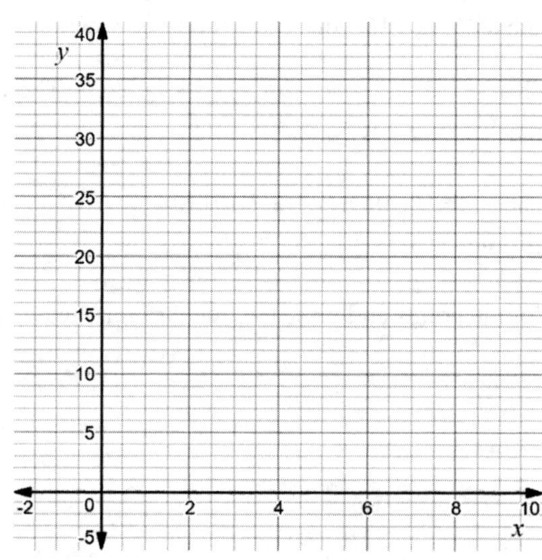

1. When $x = 4$, then $y =$ ____ .
2. When $x = 7$, then $y =$ ____ .
3. Plot the two points found in Exercises 1 and 2 .
4. Draw the line through the two points.

In Exercises 5–8, use the given graph and the equation $y = 0.247x + 12.93$.

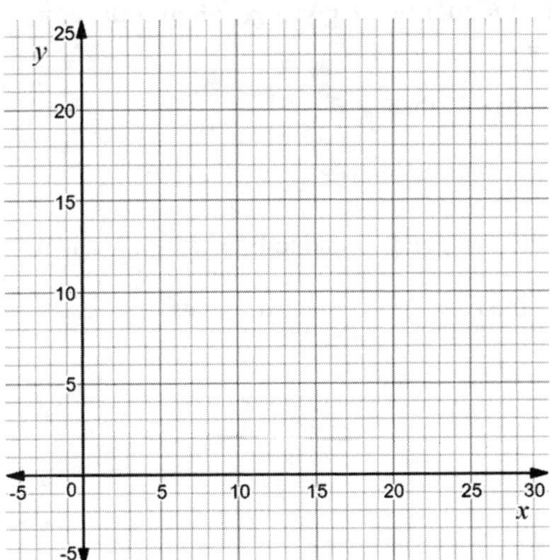

5. When $x = 10$, then $y =$ ____ .

6. When $x = 20$, then $y =$ ____ .

7. Plot the two points found in Exercises 5 and 6 .

8. Draw the line through the two points.

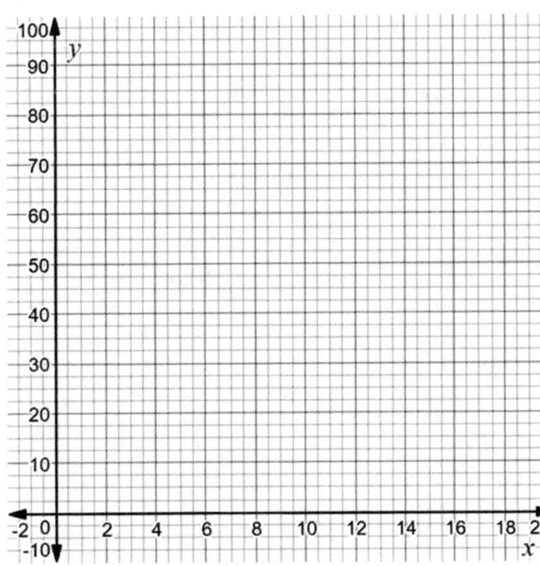

In Exercises 9–12, use the given graph and the equation $y = 5.16x + 1.3$.

9. When $x = 6$, then $y =$ ____ .

10. When $x = 10$, then $y =$ ____ .

11. Plot the two points found in Exercises 9 and 10 .

12. Draw the line through the two points.

In Exercises 13–16, use the given graph and the equation $y = -130.45x + 850$.

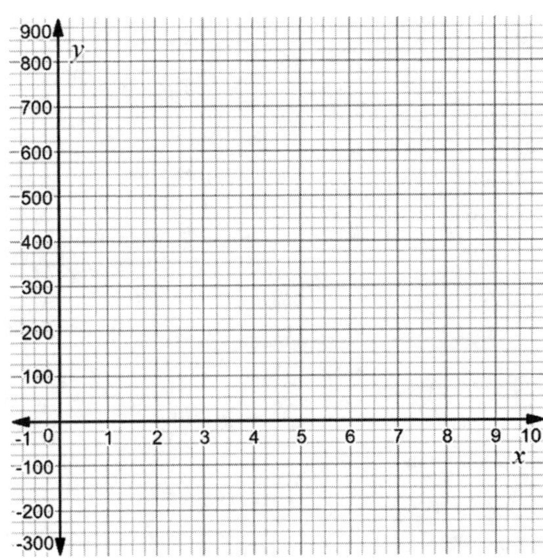

13. When $x = 4$, then $y =$ ____ .

14. When $x = 7$, then $y =$ ____ .

15. Plot the two points found in Exercises 13 and 14 .

16. Draw the line through the two points.

In Exercises 17–20, use the given graph and the equation $y = -0.24x + 7.50$.

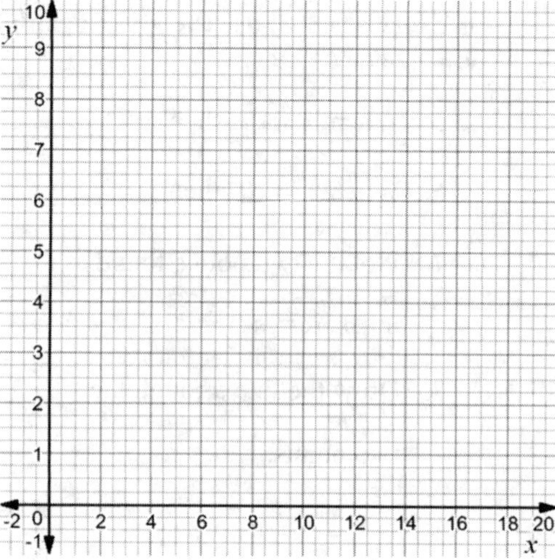

17. When $x = 8$, then $y =$ ____ .

18. When $x = 15$, then $y =$ ____ .

19. Plot the two points found in Exercises 17 and 18 .

20. Draw the line through the two points.

Answer Key

Reading Graphs (Section 1.1)

1. 20 customers
2. 15 customers
3. Week 1
4. Store B
5. *Sample answer:* Both Store A and Store B had their maximum number of new customers in Week 1.
6. 100 customers
7. 120 customers
8. 2000 calories
9. about 50%
10. Dinner
11. 11.25%
12. 68.75%
13. Lunch
14. 19 students
15. 5 students
16. Silver
17. Both Ms. T's Class and Mrs. P's Class
18. Mr K's Class
19. 60 medals
20. Ms. T's Class; platinum

Basic Mathematical Symbols and Greek Letters (Section 1.1)

1. False. It is important to distinguish between uppercase and lowercase, and to keep the original form.
2. False. Greater than does not include equality.
3. False. $x = \pm 7$ is the same as $x = 7$ or $x = -7$.
4. True. A rounded answer is an approximate answer.
5. True by the definitions of $<$ and $\leq$.
6. False. Using one letter with subscripts is a better choice.
7. $\bar{x}$, x-bar (repeated 5 times).
8. $\hat{p}$, p-hat (repeated 5 times).
9. α, alpha (repeated 5 times).
10. μ, mu (repeated 5 times).
11. σ, lowercase sigma (repeated 5 times).
12. χ, chi (repeated 5 times).
13. Σ, uppercase sigma (repeated 5 times).
14. ρ, rho (repeated 5 times).
15. All values less than but not including 1.
16. All values greater than and includes 1.
17. Any value except 1 (includes less than and greater than).
18. 1.
19. A value close to 1, but not 1.
20. 1 and -1.

Percents (Section 1.1)

1. $\frac{37}{50}$
2. $\frac{37}{200}$
3. $\frac{2}{5}$
4. $\frac{427}{500}$
5. $\frac{49}{400}$
6. $\frac{1}{200}$
7. 0.43
8. 0.89145
9. 0.2206
10. 0.390021
11. 0.0634
12. 0.0057
13. 61%
14. 41%
15. 35.85%
16. 29.74%
17. 4.9%

18. 15.38%

19. The population proportion p of members of the XYZ Fitness Center who eat yogurt at least five days a week is between 54.55% and 72.73%.

20. The population proportion p of cats who like watching TV is between 65.22% and 78.26%.

Addition and Subtraction of Integers (Section 1.2)

1. –12

2. 5

3. –9

4. –44

5. 24 – 23

6. 45 – 65

7. 116 – 52

8. 188 – 383

9. 359 – 451

10. 2000 – 2500

11. –19

12. –35

13. –62

14. –37

15. –110

16. 248

17. –65

18. –53

19. 600

20. –1519

Multiplication and Division of Integers (Section 2.1)

1. –32

2. 0

3. 63

4. 210

5. 72

6. 900

7. –88

8. –120

9. –720

10. –2,646,000

11. –7

12. 0

13. 8

14. Undefined

15. –2

16. –4

17. 0

18. 20

19. Undefined

20. $\frac{9}{0} = 0$ is not a true statement because $9 \neq 0 \cdot 0$.

Order of Operations (Section 2.1)

1. 38

2. –6

3. –29

4. 38

5. 11

6. 53

7. 17

8. –14

9. 16

10. 58

11. 17

12. –11

13. 6

14. 35

15. 18

16. 1

17. 48

18. 0

19. 1

20. 35

Operations with Fractions and Decimals (Section 2.1)

1. $\frac{3}{2}$

2. $\frac{3}{5}$

3. $\frac{4}{11}$

4. $\frac{7}{5}$

5. $\frac{4}{15}$

6. $\frac{14}{5}$

7. 1

8. $\frac{24}{5}$

9. $\frac{10}{9}$

10. $\frac{1}{7}$

11. $\frac{8}{65}$

12. $\frac{88}{27}$

13. $\frac{1}{10}$

14. $\frac{19}{12}$

15. $\frac{41}{40}$

16. $\frac{16}{63}$

17. 0.87

18. 0.47

19. 0.60

20. 0.95

Significant Digits and Rounding (Section 2.1)

1. 4
2. 2
3. 1 or 2
4. 2, 3, 4, or 5
5. 5
6. 5
7. 6
8. 6
9. 6
10. 4
11. 0.288
12. 5,210
13. 12.3
14. 0.00942
15. 68,000
16. 54,100
17. 3.14
18. 0.0780
19. 155
20. 0.0123

The Number Line and Ordering Numbers (Section 2.2)

1. −3 −2 −1 0 1 2 3

2. 30 31 32 33 34 35 36 37 38

3. 62 63 64 65 66 67 68 69 70 71 72

4. −6 −5 −4 −3 −2 −1 0

5. 0 2 4 6 8 10 12 14 16

6. −25 −20 −15 −10 −5 0 5 10 15 20 25

7. 0 0.5 1 1.5 2 2.5 3 3.5 4

8. 10 10.1 10.2 10.3 10.4 10.5

9. 0 0.25 0.5 0.75 1

10. $-1 \quad -\frac{3}{4} \quad -\frac{1}{2} \quad -\frac{1}{4} \quad 0$

11. $-14, -6, -3, 7, 9, 11, 12$

12. $-16, -14, -10, -5, 0, 8, 14, 20$

13. $-61, -52, -35, -26, 18, 24, 29, 40, 50$

14. $-114, -96, -73, 39, 75, 104, 121$

15. $-375, -350, -321, -319, -314, -296, -273$

16. $-3.5, -1.5, -1.25, -1, 1.5, 2.25, 2.75, 3.25$

17. $-8.5, -8.1, -7.9, -6.3, 5.8, 6.5, 7.4, 7.5$

18. $-92, -87.58, -81, -65.15, -54.10, 75.18, 79, 85.23$

19. $-\frac{7}{12}, -\frac{5}{12}, -\frac{1}{12}, \frac{1}{12}, \frac{5}{12}, \frac{7}{12}$

20. $\frac{1}{12}, \frac{1}{4}, \frac{5}{12}, \frac{1}{2}, \frac{7}{12}, \frac{3}{4}, \frac{5}{6}$

The *xy*-Plane and Point Plotting (Section 2.1)

1. *x*-axis

2. *y*-axis

3. Origin; $(0, 0)$

4. Quadrants

5. 6

6. 18

7. The coordinates must be enclosed within parentheses; $(9, 1)$.

8. Bottom

9. Right

10. True

11. False

Exercises 12–20.

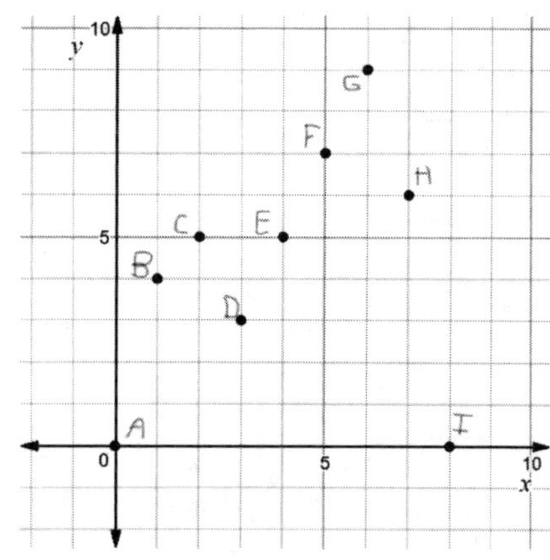

Summation (Section 2.1)

1. 37

2. 51

3. 35

4. –4

5. 20.5

6. –0.5

7. –2.6

8. –81.8

9. 305

10. 181

11. –27

12. 1.5

13. 176

14. 84.5

15. 25.62

16. 26.5525

17. 66

18. 35

19. 0

20. –5

Evaluating Formulas (Section 2.1)

1. $p = \frac{\mu}{n}$

2. $w = \frac{P - 2l}{2}$ or $w = \frac{P}{2} - l$

3. $x = z\sigma + \mu$

4. $N = \frac{\Sigma x}{\mu}$

5. $\frac{np}{100} = f$

6. $k = \frac{y + M}{b}$

7. $n = \frac{5r + T}{T}$ or $n = \frac{5r}{T} + 1$

8. 88

9. 5

10. –3.2

11. –13.5

12. 80

13. 1.6

14. 3.2

15. 7.6

16. –1.3

17. 0.68

18. –20

19. 17

20. 25

Operations with Summations (Section 2.3)

1. 10.5

2. 0.7

3. 0

4. –70.5

5. 154.7

6. 0

7. 700

8. –473

9. –1106.7

10. –34.2

11. –0.5

12. –2.9

13. –199.3

14. –36.5

15. 5.0

16. 5

17. 77.1

18. 0.17

19. 0.34

20. –0.01

Exponents (Section 2.4)

1. 81

2. –100

3. –16

4. –1

5. –36

6. 169

7. $-\frac{1}{25}$

8. $\frac{1}{4}$

9. 16

10. 64

11. 25

12. 25

13. $-\frac{1}{3}$

14. $-\frac{20}{3}$

15. $\frac{5}{18}$

16. –36

17. 8

18. $\frac{9}{16}$

19. –1.08

20. 24.7

Operations with Square Roots (Section 2.4)

1. 11

2. 0

3. –12

4. Not a real number

5. 6

6. 10

7. 4

8. $\frac{1}{2}$

9. $3\sqrt{7}$

10. $10\sqrt{2}$

11. $4\sqrt{5}$

12. $6\sqrt{3}$

13. $3\sqrt{10}$

14. $5\sqrt{6}$

15. $7\sqrt{3}$

16. $2\sqrt{2}$

17. $\frac{\sqrt{3}}{3}$

18. $\frac{\sqrt{10}}{5}$

19. $\dfrac{2\sqrt{15}}{5}$

20. $\dfrac{\sqrt{2}}{2}$

Division Involving Powers and Square Roots (Section 2.4)

1. (a) $\dfrac{\sqrt{61.6}}{5}$

   (b) 1.57

2. (a) $\dfrac{\sqrt{36.95}}{5}$

   (b) 1.22

3. (a) $\dfrac{\sqrt{225.68}}{8}$

   (b) 1.88

4. (a) $\dfrac{\sqrt{56.3}}{5}$

   (b) 1.50

5. (a) $\dfrac{\sqrt{47.1}}{5}$

   (b) 1.37

6. (a) $\dfrac{\sqrt{91.98}}{7}$

   (b) 1.37

7. (a) $\dfrac{\sqrt{33.92}}{8}$

   (b) 0.73

8. (a) $\dfrac{\sqrt{98.32}}{8}$

   (b) 1.24

9. 1.02
10. 0.98
11. 1.13
12. 0.70
13. 1.13
14. 1.07
15. 0.54
16. 3.80
17. 320.07
18. 12,360.13
19. 44.91
20. 1220.08

Absolute Value (Section 2.4)

1. 12
2. 4.5
3. 0.27
4. –0.051
5. –0.94
6. 0
7. 0.5
8. –0.5
9. 0.2
10. –0.2025
11. 0.05
12. –8
13. 4.37
14. –3.6426
15. 0.17
16. 0.16
17. $\dfrac{1}{2}$
18. 2.9
19. 0.738
20. 2.4

More Multiplication and Division (Section 3.1)

1. $1^3 \cdot 4 \cdot 5^5$
2. $3^2 \cdot 4^3 \cdot 6 \cdot 9^3$
3. $5^6 \cdot 11^3$
4. $3 \cdot 6^3 \cdot 10 \cdot 13^5 \cdot 17^2$
5. $12^4 \cdot 15^2 \cdot 20^3 \cdot 21 \cdot 25 \cdot 27$
6. $4^7 \cdot 5 \cdot 9^2 \cdot 11^5$
7. $56^3 \cdot 62^4 \cdot 63^2 \cdot 70^5$
8. $100^3 \cdot 101^4 \cdot 102^4 \cdot 103 \cdot 104^7$
9. $p = 0.625$

10. $r \approx 0.15$

11. $t \approx 0.17$

12. $c = 1.8$

13. $w \approx 12.67$

14. $p = 0.02$

15. $s = 0.019$

16. $h = 0.00154$

17. $v \approx 0.29$

18. $q \approx 0.02$

19. $b \approx 0.47$

20. $r = 0.8125$

Double Inequalities (Section 3.1)

1. y: −5 −4 −3 −2 −1 0 1 2 3

2. w: −1 0 1 2 3 4 5 6 7

3. t: −3 −2 −1 0 1 2 3 4 5

4. z: 0 1 2 3 4 5 6 7 8

5. P: −3 −2 −1 0 1 2 3 4 5

6. x: −2 −1 0 1 2

7. z: −3 −2 −1 0 1

8. y: 0 1 2 3 4

9. $-6 \le x \le 2$

10. $8 \le x < 15$

11. $9 \le x \le 29$

12. $-0.14 \le x \le 0.34$

13. $-4.22 < x < 0.72$

14. $-2 < x < 6$

15. $-8 \le z \le 5$

16. $2 < z < 10$

17. $-5 \le x \le 7$

18. $-19 < x < 13$

19. $0.5 \le x \le 5.5$

20. $-2.5 \le x \le 6.5$

Translating Verbal Phrases (Section 3.2)

1. $12 < d < 18$

2. $w \le 55$

3. $m = 5$

4. $s \ge 100$

5. $g \ne 0$

6. $t > 80$

7. $p \approx 500$

8. $g < 3$

9. $16 \le c \le 78$

10. $s \le 3$

11. $w \approx 0.32$

12. $p > 50$

13. $P \ne 9$

14. $150 < g < 200$

15. $c = 500$

16. $t \ge 15$

17. $1600 \le s \le 2500$

18. $h < 12$

19. $d \approx 76.28$

20. $p \le 0.18$

Factorials (Section 3.4)

1. 6

2. 120

3. 40,320

4. 1

5. 362,880

6. 720

7. 1

8. 479,001,600

9. 1

10. 6,227,020,800

11. 120

12. 1

13. 6,227,020,800
14. 39,916,800
15. 362,880
16. 479,001,600
17. 1
18. 120
19. 720
20. 6,227,020,800

Operations with Factorials (Section 3.4)

1. 6
2. 240
3. 576
4. 1440
5. 4
6. $\frac{1}{336}$
7. 7
8. 10,626
9. 13
10. 73,440
11. 726,485,760
12. $\frac{1}{36}$
13. 360
14. 24
15. 27,132
16. 6
17. 2300
18. 70
19. 2002
20. $\frac{1}{18}$

Scientific Notation (Section 3.4)

1. 6.218547×10^{8}, large
2. $7.3900045 \times 10^{-12}$, small
3. 5×10^{-14}, small
4. 1.2500649×10^{22}, large
5. 3.995475×10^{18}, large
6. -6.425137×10^{27}, large
7. $-9.5000024 \times 10^{-17}$, small
8. 2.0067818×10^{21}, large
9. -3×10^{24}, large
10. $8.2384579 \times 10^{-31}$, small
11. 3.82714×10^{-9}, 3.83×10^{-9}
12. $5.0071824 \times 10^{-18}$, 5.01×10^{-18}
13. 9.3627121×10^{17}, 9.36×10^{17}
14. $6.00031892 \times 10^{13}$, 6.00×10^{13}
15. $4.1182647 \times 10^{-25}$, 4.12×10^{-25}
16. $7.1553454 \times 10^{-10}$, 7.16×10^{-10}
17. $8.88964132 \times 10^{25}$, 8.89×10^{25}
18. $1.91842374 \times 10^{37}$, 1.92×10^{37}
19. $2.0032598 \times 10^{-31}$, 2.00×10^{-31}
20. $8.44494287 \times 10^{28}$, 8.44×10^{28}

More Translating Verbal Phrases (Section 4.2)

1. $n = 0, 1, 2, 3$
2. $n = 0, 1, 2, 3, 4, 5, 6$
3. $n = 5, 6, 7, 8$
4. $n = 10, 11, 12, 13, \ldots$
5. $n = 3, 4, 5, 6$
6. $n = 0, 1, 2$
7. $n = 6, 7, 8$
8. $n = 4, 5, 6, 7, 8$
9. $n = 7, 8, 9, 10, \ldots$
10. $n = 1, 2, 3, 4, 5, 6, 7, 8, 9$
11. $n = 1, 2, 3, 4, 5$
12. $n = 0, 1, 2, 3, 4, 5, 6$
13. $n = 3, 4, 5, 6, 7, 8, 9, 10$
14. $n = 4, 5, 6, 7$
15. $n = 21, 22, 23, 24, 25$
16. $n = 3, 4$
17. $n = 0, 1$
18. $n = 0, 1, 2, 3, 4, 5, 6, 7$

19. $n = 50, 51, 52, 53, \ldots$

20. $n = 0, 1, 2$

More Division Involving Square Roots (Section 5.1)

1. 0.61
2. 1.34
3. 0.30
4. 1.83
5. 1.45
6. 1.82
7. 0.82
8. 0.45
9. 0.56
10. 0.38
11. 1.70
12. 1.20
13. 1.26
14. 3.05
15. 2.32
16. 0.56
17. 2.74
18. 0.54
19. 3.06
20. 0.71

Intervals on the Number Line (Section 6.1)

1.

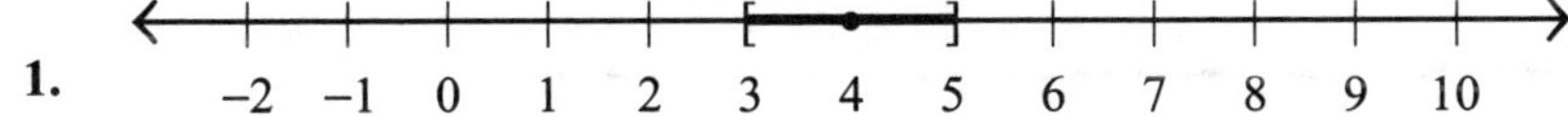

2. –6 –5 –4 –3 –2 –1 0 1 2 3 4 5 6

3. –8 –7 –6 –5 –4 –3 –2 –1 0 1 2 3 4

4. –50 –40 –30 –20 –10 0 10 20 30 40 50 60 70

5. 0 50 100 150 200 250 300 350 400 450 500 550 600

6. –3 –2.5 –2 –1.5 –1 –0.5 0 0.5 1 1.5 2 2.5 3

7. 0 0.5 1 1.5 2 2.5 3 3.5 4 4.5 5 5.5 6

8. –4 –3.5 –3 –2.5 –2 –1.5 –1 –0.5 0 0.5 1 1.5 2

9. –1.5 –1 –0.5 0 0.5 1 1.5 2 2.5 3 3.5 4 4.5

10. –4.5 –4 –3.5 –3 –2.5 –2 –1.5 –1 –0.5 0 0.5 1 1.5

11.

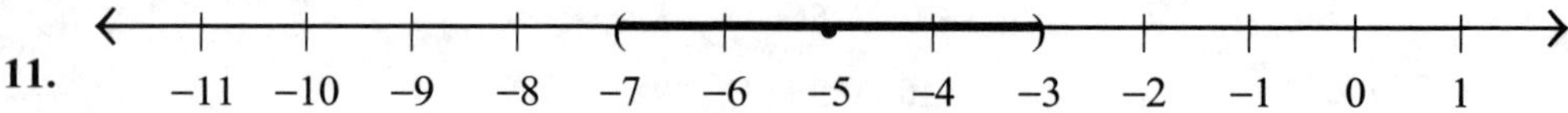

12.

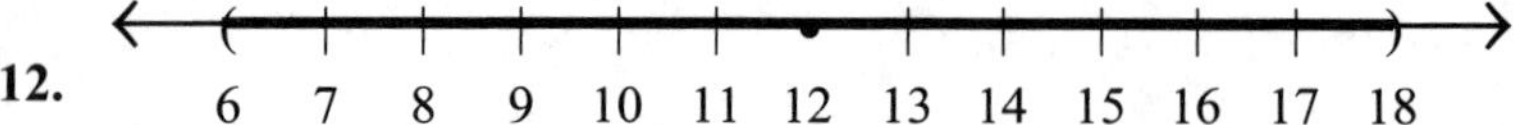

13.

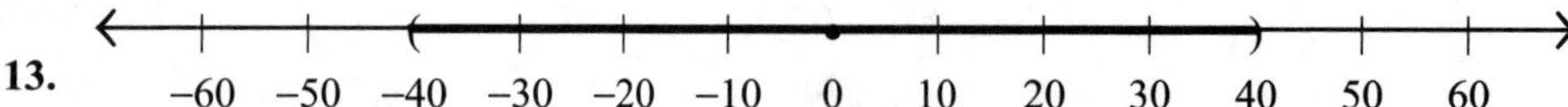

14.

−80 −70 −60 −50 −40 −30 −20 −10 0 10 20 30 40

15.

16.

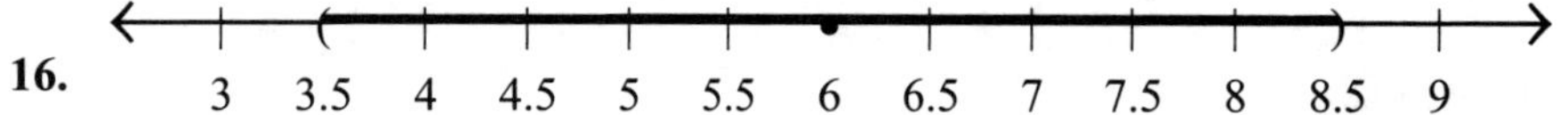

17.

−9 −7.5 −6 −4.5 −3 −1.5 0 1.5 3 4.5 6 7.5 9

18.

−6 −5.5 −5 −4.5 −4 −3.5 −3 −2.5 −2 −1.5 −1 −0.5 0

19.

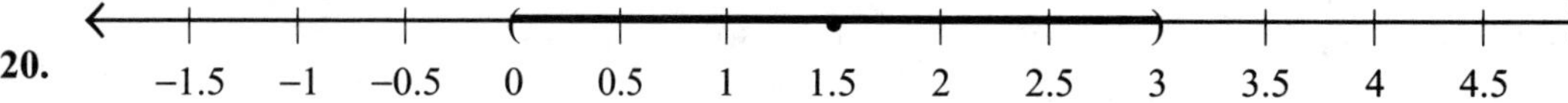

20.

−1.5 −1 −0.5 0 0.5 1 1.5 2 2.5 3 3.5 4 4.5

Slope-Intercept Form of the Equation of a Line (Section 9.2)

1. 12.78

2. 23.73

3–4.

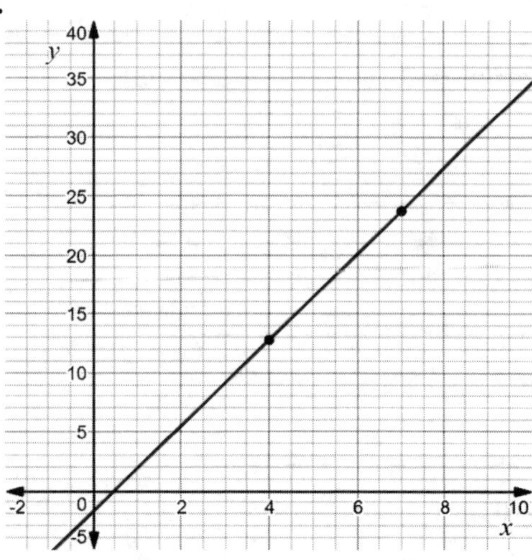

5. 15.4
6. 17.87
7–8.

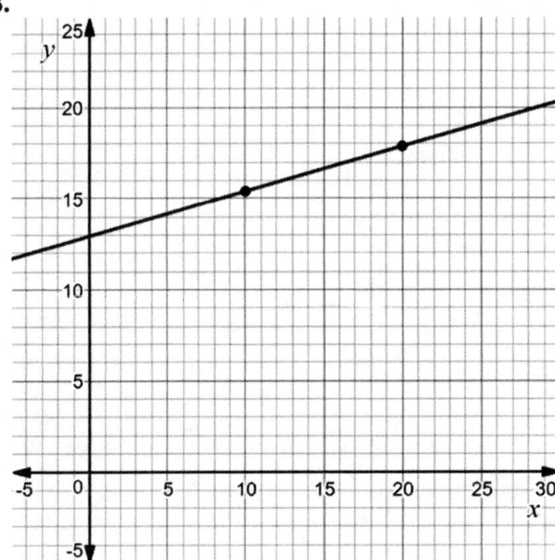

9. 32.26
10. 52.9
11–12.

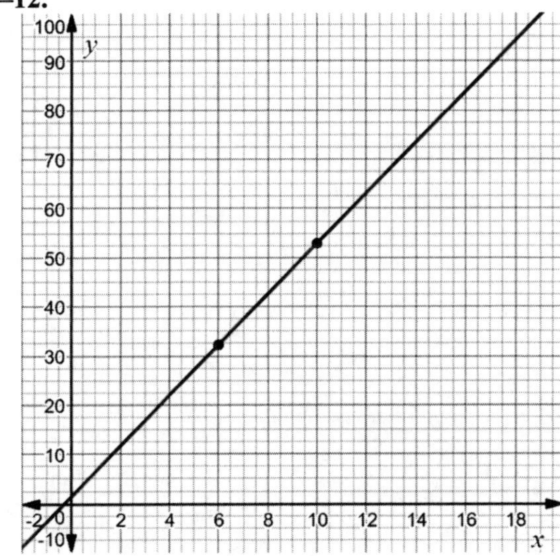

13. 328.2
14. –63.15

15–16.

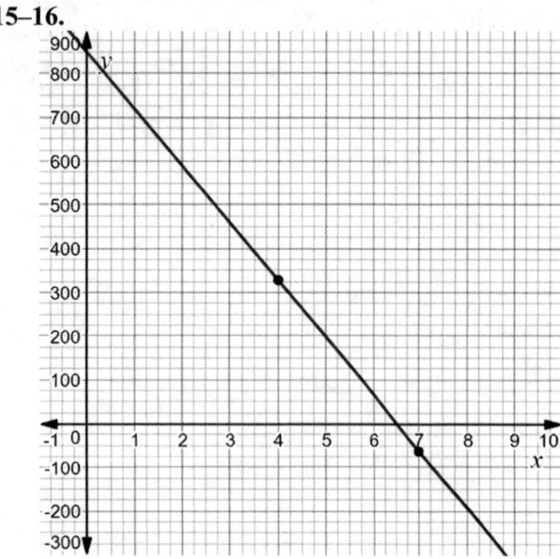

17. 5.58

18. 3.9

19–20.

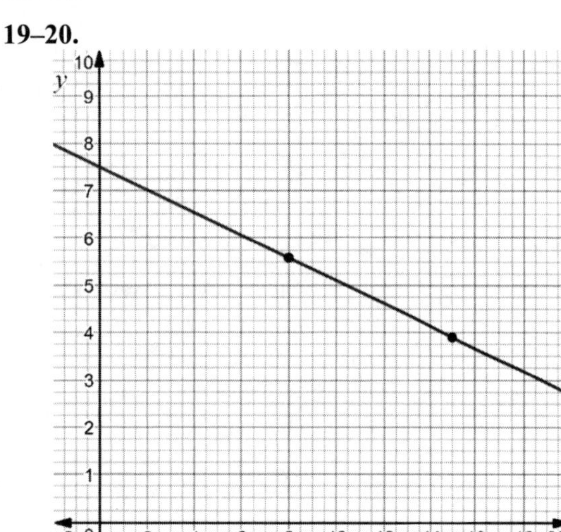